Lecture Notes in Electrical Engineering

Volume 329

About this Series

"Lecture Notes in Electrical Engineering (LNEE)" is a book series which reports the latest research and developments in Electrical Engineering, namely:

- Communication, Networks, and Information Theory
- Computer Engineering
- Signal, Image, Speech and Information Processing
- Circuits and Systems
- Bioengineering

LNEE publishes authored monographs and contributed volumes which present cutting edge research information as well as new perspectives on classical fields, while maintaining Springer's high standards of academic excellence. Also considered for publication are lecture materials, proceedings, and other related materials of exceptionally high quality and interest. The subject matter should be original and timely, reporting the latest research and developments in all areas of electrical engineering.

The audience for the books in LNEE consists of advanced level students, researchers, and industry professionals working at the forefront of their fields. Much like Springer's other Lecture Notes series, LNEE will be distributed through Springer's print and electronic publishing channels.

More information about this series at http://www.springer.com/series/7818

James J. (Jong Hyuk) Park · Yi Pan
Cheonshik Kim · Yun Yang
Editors

Future Information Technology - II

Springer

Editors
James J. (Jong Hyuk) Park
Seoul National University of Science and Technology (SeoulTech)
Seoul
Korea, Republic of (South Korea)

Cheonshik Kim
Digital Media Engineering
Anyang University
Anyang
Korea, Republic of (South Korea)

Yi Pan
Department of Computer Science
Georgia State University
Atlanta, GA
USA

Yun Yang
Information and Communication Technologies
Swinburne University of Technology
Melbourne, VIC
Australia

ISSN 1876-1100 ISSN 1876-1119 (electronic)
Lecture Notes in Electrical Engineering
ISBN 978-94-024-0369-5 ISBN 978-94-017-9558-6 (eBook)
DOI 10.1007/978-94-017-9558-6

Springer Dordrecht Heidelberg New York London

Softcover reprint of the hardcover 1st edition 2015

Printed on acid-free paper

Springer Science+Business Media B.V. Dordrecht is part of Springer Science+Business Media (www.springer.com)

Message from the FutureTech 2014 General Chairs

FutureTech 2014 is the FTRA nineth event of the series of international scientific conference. This conference took place from May 28–31, 2014, in Zhangjiajie China. The aim of the FutureTech 2014 was to provide an international forum for scientific research in the technologies and application of information technologies. It was organized by the Korea Information Technology Convergence Society.

FutureTech 2014 is the next edition of FutureTech 2013 (Gwangju, Korea), FutureTech 2012 (Vancouver, Canada), FutureTech 2011 (Loutraki, Greece), FutureTech 2010 (Busan, Korea, May 2010), which was the next event in a series of highly successful the International Symposium on Ubiquitous Applications and Security Services (UASS-09, USA, January 2009), previously held as UASS-08 (Okinawa, Japan, March 2008), UASS-07 (Kuala Lumpur, Malaysia, August, 2007), and UASS-06 (Glasgow, Scotland, UK, May 2006).

The papers included in the proceedings cover the following topics: Hybrid Information Technology High Performance Computing, Cloud and Cluster Computing, Ubiquitous Networks and Wireless Communications Digital Convergence, Multimedia Convergence, Intelligent and Pervasive Applications, Security and Trust Computing, IT Management and Service Bioinformatics and Bio-Inspired Computing, Database and Data Mining, Knowledge System and Intelligent Agent, Game and Graphics Human-centric Computing and Social Networks, Advanced Mechanical Engineering, Computer Aided Machine Design, Control and Automations and Simulation. Accepted and presented papers highlight new trends and challenges of future and information technology. The presenters showed how new research could lead to novel and innovative applications. We hope you will find these results useful and inspiring for your future research.

We would like to express our sincere thanks to Steering Chairs: James J. (Jong Hyuk) Park (SeoulTech, Korea), Hamid R. Arabnia (The University of Georgia, USA). Our special thanks go to the Program Chairs: Cheonshik Kim (Anyang University, Korea), Bin Guo (Northwestern Polytechnical University, China), Xiaojun Cao (Georgia State University, USA), Muhammad Khurram Khan (King Saud University, Saudi Arabia), Shaozi Li (Xiamen University, China), Yun Yang (Swinburne University of Technology, Australia), Beihong Jin

(Chinese Academy of Sciences, China), all Program Committee members and all the additional reviewers for their valuable efforts in the review process, which helped us to guarantee the highest quality of the selected papers for the conference.

We cordially thank all the authors for their valuable contributions and the other participants of this conference. The conference would not have been possible without their support. Thanks are also due to the many experts who contributed to making the event a success.

May 2014

Yi Pan, Georgia State University, USA
XianfengLuo, Shanghai University, China
Xiaohong Jiang, FUN, Hakodate, Japan
Young-Sik Jeong, Dongguk University, Korea

Message from the FutureTech 2014 Program Chairs

Welcome to the FTRA Nineth FTRA International Conference on Future Information Technology (FutureTech 2014) held in Zhangjiajie, China during May 28–31, 2014. FutureTech 2014 is the most comprehensive conference focused on the various aspects of information technology. FutureTech 2014 provides an opportunity for academic and industry professionals to discuss recent progress in the area of information technology. In addition, the conference will publish high quality papers, which are closely related to the various theories and practical applications in information technology. Furthermore, we expect that the conference and its publications will be a trigger for further related research and technology improvements in these important subjects.

For FutureTech 2014, we received many paper submissions, after a rigorous peer review process, we accepted the articles with high quality for the FutureTech 2014 proceedings, published by the Springer. All submitted papers have undergone blind reviews by at least three reviewers from the technical program committee, which consists of leading researchers around the globe. Without their hard work, achieving such a high-quality proceeding would not have been possible. We take this opportunity to thank them for their great support and cooperation. We would like to sincerely thank the following invited speakers who kindly accepted our invitations, and, in this way, helped to meet the objectives of the conference: Prof. Han-Chieh Chao (National Ilan University, Taiwan) and Timothy K. Shih (National Central University, Taiwan). Finally, we would like to thank all of you for your participation in our conference, and also thank all the authors, reviewers, and organizing committee members. Thank you and enjoy the conference!

Cheonshik Kim, Anyang University, Korea
Bin Guo, Northwestern Polytechnical University, China
Xiaojun Cao, Georgia State University, USA
Muhammad Khurram Khan, King Saud University, Saudi Arabia
Shaozi Li, Xiamen University, China
Yun Yang, Swinburne University of Technology, Melbourne, Australia
Beihong Jin, Professor, Chinese Academy of Sciences, China

Organization

Steering Chairs

James J. Park, SeoulTech, Korea
Hamid R. Arabnia, The University of Georgia, USA

General Chairs

Yi Pan, Georgia State University, USA
XianfengLuo, Shanghai University, China
Xiaohong Jiang, FUN, Hakodate, Japan
Young-Sik Jeong, Dongguk University, Korea

Program Chairs

Cheonshik Kim, Anyang University, Korea (Leading Chair)
Bin Guo, Northwestern Polytechnical University, China
Xiaojun Cao, Georgia State University, USA
Muhammad Khurram Khan, King Saud University, Saudi Arabia
Shaozi Li, Xiamen University, China
Yun Yang, Swinburne University of Technology, Melbourne, Australia
Beihong Jin, Professor, Chinese Academy of Sciences, China

Workshop Chairs

Young-Gab Kim, Korea University, Korea
Lei Ye, University of Wollongong, Australia
Hiroaki Nishino, Oita University, Japan
Neil Y. Yen, The University of Aizu, Japan

Publication Chair

Sung-Ki Kim, Sun Moon University, Korea (Leading Chair)
Jing (Selena) He, Kennesaw State University, USA
Haixia Zhang, Shandong University, China
Zonghua Zhang, Institut TELECOM/TELECOM Lille1, France
Arthur H. Yang, Tungnan University, Taiwan
Xu Shuo, ISTIC, China
Vincent Huang, National Taichung University of Science and Technology, Taiwan

International Advisory Committee

Qun Jin, Waseda, Japan
Fatos Xhafa, Technical University of Catalonia, Spain
Hsiao-Hwa Chen, National Cheng Kung University, Taiwan
Ivan Stojmenovic, University of Ottawa, Canada
Doo-soon Park, SoonChunHyang University, Korea
Makoto Takizawa, Seikei University, Japan
Mohammad S. Obaidat, Monmouth University, USA
Han-Chieh, Chao National Ilan University, Taiwan
Laurence T. Yang, St. Francis Xavier University, Canada
Young-Sik Jeong, Wonkwang University, Korea
Albert Zomaya, University of Sydney, Australia
Wanlei Zhou, Deakin University, Australia
Ruppa K. Thulasiram, University of Manitoba, Canada
Hai Jiang, Arkansas State University, USA
Yuanchun Shi, Tsinghua University, China

Publicity Chairs

Byung-Gyu Kim, Sun Moon University, Korea
Yu Chen, State University of New York, Binghamton, USA
Yun Cheng, Hunan University of Humanities, Science and Technology, China
Masaru Fukushi, Yamaguchi University, Japan
Wei Zhong, University of South Carolina Upstate, USA
Maiga Chang, Athabasca University, Canada
Chun-Hong Huang, Lunghwa University, Taiwan
Hui-Kai Su, National Formosa University, Taiwan
OscalTzyh-Chiang Chen, National Chung Cheng University, Taiwan
Kuo-Chung Chu, National Taipei University and Health Sciences, Taiwan

Gangman Yi, Gangneung-Wonju National University, Korea
Anna Kang, Dongguk University, Korea
Namje Park, Jeju National University, Korea

Invited Speaker

Han-Chieh Chao, National Ilan University, Taiwan
Timothy K. Shih, National Central University, Taiwan

Program Committee

Abdelbadeeh Salem, Ain Shams University, Egypt
Alfredo Cuzzocrea, University of Calabria, Italy
Andre Valdestilhas, Brazil Research Group, Brazil
Ashkan Sami, Shiraz University, Iran
Carson Leung, University of Manitoba, Canada
Cheng Zhang, Anhui University, China
Chi-Fu Huang, National Chung Cheng University, Taiwan
Ching-Hsien Hsu, Chung Hua University, China
Cliff Zou, University of Central Florida, USA
Colin Walter, University of London, UK
Dong Yuan, Swinburne University of Technology, Australia
Epaminondas Kapetanios, University of Westminster, UK
Guillermo Diaz Diaz-Delgado, Universidad Autónoma de Querétaro, Mexico
Hariharan Shanmugasundaram, TRP Engineering College, India
Hirohi Ishikawa, Tokyo Metropolitan University, Japan
Hongmei Chi, Florida Agricultural and Mechanical University, USA
Hongyu Zhang, Tsinghua University, China
Jiehan Zhou, Carleton University, Canada
Jinli Cao, LaTrobe University, Australia
Jiqiang Lu, Institute for Infocomm Research, Singapore
Jun Yan, University of Wollongong, Australia
Junchao Xiao, Academy of Sciences, China
Kuo-Chan Huang, National Taichung University of Education, Taiwan
Kyungbaek Kim, Chonnam National University, Korea
Li Li, Southwest University, China
Lizhen Cui, Shandong University, China
Maumita Bhattacharya, Charles Sturt University, Australia
Mei-Ling Shyu, University of Miami, USA
Mudasser Wyne, National University, USA
Muhammad Usman, Auckland University of Technology, New Zealand
Norman Saeed, LaTrobe University, Australia
Pai-Ling Chang, Shih-Hsin University, Taiwan

Paul Kwan, University of New England, Australia
Qiang He, Swinburne University of Technology, Australia
Qing Yang, Montana State University, USA
Ren-Song Ko, National Chung Cheng University, Korea
Ruben Rios, Universidad de Malaga, Spain
Ryszard Tadeusiewicz, AGH University of Science and Technology, Poland
Shu-Ching Chen, Florida International University, USA
Song Fu, University of North Texas, USA
Soon M. Chung, Wright State University, USA
Sylvia Encheva, Stord/Haugesund University College, Norway
Tatjana Davidovic, Mathematical Institute of the Serbian Academy of Sciences and Arts, Serbia
Toshiyuki Amagasa, University of Tsukuba, Japan
Vitaly Klyuev, University of Aizu, Japan
Vitomir Kovanovic, Simon Fraser University, Canada
Wei-Chuen Yau, Multimedia University, Malaysia
Willy Picard, Poznan University of Economics, Poland
Wladyslaw Homenda, Instytut Badań Systemowych Polskiej Akademii Nauk, Poland
Xiao Liu, East China Normal University, China
Yo-Ping Huang, National Taipei University of Technology, Taiwan
Yu-Chen Hu, Providence University, Taiwan
Yunquan Zhang, Institute of Computing Technology, CAS, China
Zeyar Aung, Masdar Institute, UAE

Contents

Chapter 1
10-Step Information Architecture on Wanfang Med Online

Xiumei Zhang, Xiaoying Zhou, Junli Liu and Jianwu Xu

Abstract The paper first reviews the different definitions and the basic fundamentals of information architecture, then summarizes the creative-application experiences of IA in the field of Chinese medical-information service based on the practice of Wanfang Med Online and gives out the ten steps of the IA process, which include research on users, content analysis, context analysis, metadata framework construction, classification, thesaurus, IA strategy, Project Management (PM), blueprint, and user interface.

Keywords Information architecture · Case study · Wanfang Med Online · Information architecture process

1.1 Introduction

As environmental pollution continues and society ages, more than a billion Chinese people have demonstrated strong demand for medical and health care information. Despite a wide variety of health websites, few medical-information service

---

Project 71133006 supported by National Natural Science Foundation of China; Project 71073168 supported by National Natural Science Foundation of China.

X. Zhang (✉) · J. Liu · J. Xu
Institute of Scientific and Technical Information of China, Beijing, China
e-mail: xiumei@wanfangdata.com.cn

J. Liu
e-mail: Liujl2011@istic.ac.cn

J. Xu
e-mail: xujianwu@wanfangdata.com.cn

X. Zhang · X. Zhou
Renmin University of China, Beijing, China
e-mail: xyzhou@263.net

J.J. (Jong Hyuk) Park et al. (eds.), *Future Information Technology - II*,
Lecture Notes in Electrical Engineering 329,
DOI 10.1007/978-94-017-9558-6_1

platforms are actually scientific, authoritative, and trustworthy. The library and informatics field (LIS) has accumulated a wealth of information architecture (IA) theoretical achievements. The research on how to combine these results with the Chinese medical information service has far-reaching significance, on the one hand expanding and promoting IA application, and on the other hand exploring the design theory of medical-information service.

1.2 Information Architecture and the Practice of Wanfang Med Online

1.2.1 Three Classical Definitions and Four Basic Fundamentals of Information Architecture

1.2.1.1 Three Classical Definitions of Information Architecture

Wurman defines a founder of IA in these terms: (1) a person who organizes the patterns inherent in data, making the complex clear; (2) a person who creates the structure of an information map that allows others to find their personal paths to knowledge; (3) an emerging 21st-century professional occupation addressing the needs of the age focused upon clarity, human understanding, and the science of the organization of information [5].

Ding and Lin [1] give a definition in Information Architecture: information architecture is about organizing and simplifying information design, integrating and aggregating information spaces (systems); creating ways for people to find, understanding, exchange and manage information; and, therefore, staying on top of information and making the right decisions.

Professor Zhou [7] in China also defines IA as follows: an art and a science, which organizes information and designs information environments, information spaces, and information structure to meet the information needs of users and to help them meet their objectives.

All above three definitions of IA stress the organization and design of information, but they are different just in the form of expression.

1.2.1.2 Four Basic Fundamentals of Information Architecture [2]

- **Integration principle of information segment**: IA process is the process of integrating harmoniously and comprehensively collected information with various media from multiple aspects since the time of the collection of each information segment. There are three most important problems in the integrated process: the first is to harmoniously integrate information resources; the second is to effectively use different communication media or tools to realize mutual compatibility; and the third is to realize the function of integration with a small screen.

- **Ordering principle of information ensemble**: Organize information contents and express information forms in an information ensemble in order to do three things: substantially enhance effective information content, consciously control the entropy of the information structure system, and form methodical and logical information structure systems with distinct themes and clear primary and secondary relations. As the objects of IA, information carriers, forms, technological means, and other aspects are polyphyletic and complicated. The effect of IA depends mainly on the size of effective information content after the organization of information, while the size of effective information content is related closely to the ordering degree of information organization.
- **Display principle of information structure**: The information architect designs a coherent and functional information framework for information after ordering it. In this way, the information can be substantially expressed through the information interface and effectively display contents, styles, and characteristics, and the users can perceive information in the information structure and conveniently and happily obtain information from the information structure so as to meet information demands and realize their objectives.
- **Optimization principle of information space**: Help people remit psychological confusion or action dilemmas caused by information environment in complicated and huge information spaces, lighten their cognitive loads, enhance their abilities to perceive and capture information, and promote information acceptance and use.

1.3 The Practice of IA in Wanfang Med Online

1.3.1 Introduction of Wanfang Med Online

Wanfang Med Online is a large Chinese medical-information service platform which was developed by Wanfang Data Co, Ltd. Based on an exclusive strategic partnership with the Chinese Medical Association (CMA) and the Chinese Medical Doctor Association (CMDA), it has become the authoritative website for understanding Chinese medical and health information in mainland China and the Asia-Pacific region, serving thousands of medical institutions and nearly one million daily visits [3].

1.3.2 Innovation of IA in "Wanfang Med Online"

The information architecture of a typical website focuses on content, users, and background, but it is not enough, for a site's success, to pay attention to these alone [6]. Many related issues properly addressed and developed in coordination are also

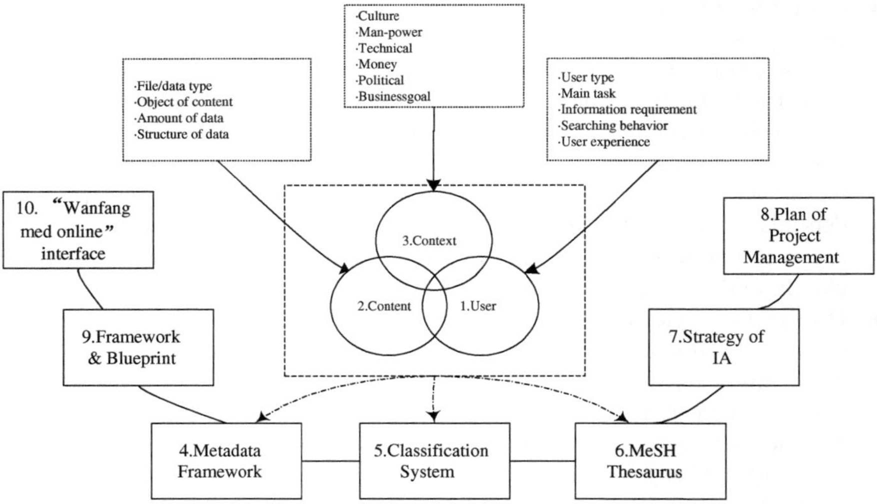

Fig. 1.1 10-step IA in Wanfang Med Online

very important. This paper proposes a 10-step website IA process applied in the construction of Wanfang Med Online, based on previous achievements described in figures or tables (Fig. 1.1).

Step 1. Research on users on Wanfang Med Online
For different types of users, Wanfang Med Online provide different solutions in main task, information requirement, searching behavior, user experience and thesaurus (Table 1.1).

Step 2. Content analysis for Wanfang Med Online
The detailed content in Wanfang Med Online is described from the following four aspects, these resources are the basis of the Follow-Up Services (Table 1.2).

Step 3. Context analysis for Wanfang Med Online
As a platform, Wanfang Med Online linked service provider and service receptor, but they have totally different goals and background. How to bridge two parties and integrated their requirements into the same website? Context analysis is very important (Table 1.3).

Step 4. Creating a metadata framework for Wanfang Med Online

- **Five-element metadata framework for bibliography resources**: Many metadata fields were indexed for each kind of literature resources, supporting the exhibition of website navigation and search results. In addition, Wanfang Med Online also proposed a five-element (author, institution, subject, keywords, and fund) metadata framework of literature resources to support different website functions (Fig. 1.2).

Table 1.1 Research on users of Wanfang Med Online

User type	Main task	Information requirement	Searching behavior	User experience	Thesaurus
Researcher or teacher	Research teaching	Academic literature	Concerning search strategy	Frontier; accurate comprehensiveness	MeSH
Doctor or nurse	Diagnosis and treatment	Medical knowledge; case reports; guideline specifications	Browse evaluation	Practicality; correlation	ICD-10 classification code for clinical- and laboratory-standard drug name
Enterprise user	Research marketing	Monitoring on adverse drug reaction behaviors of users	Regular	Timeliness; comprehensiveness	ICD-10 OTC standard drug name customized thesaurus
Patient	Health advisory	Health knowledge; treatment information	Quick view; simple query	Practicality; comprehensibility	Common names of diseases

Table 1.2 Content analysis for Wanfang Med Online

Type of content	File/data type	Object of content	Amount of data (pieces)	Structure of data
Journal	Metadata + PDF full text	Journal paper	30,000,000	Unstructured
Dissertation	Metadata + PDF full text	Dissertation	350,000	Unstructured
Conference	Metadata + PDF full text	Paper	280,000	Unstructured
Book	Metadata + PDF full text	Chapter	51,850	Unstructured
Video	Metadata + AVI or FLV	Video + multimedia	2,025 kinds; 44,196 min	Unstructured
Knowledge base	Text and picture	Knowledge point		Structured

Table 1.3 Context analysis for Wanfang Med Online

Service type	Business goal	Money	Political	Technical	Manpower	Culture
Service provider	Paid knowledge services	Research and marketing	Copyright owner	Information integration, retrieval, and services	Data collection, product research and development, marketing and sales	Internet and software
Service receptor	Access to knowledge	Research funding	Purchasing the right to use	Simple information retrieval	Doctor and patient	Medical services

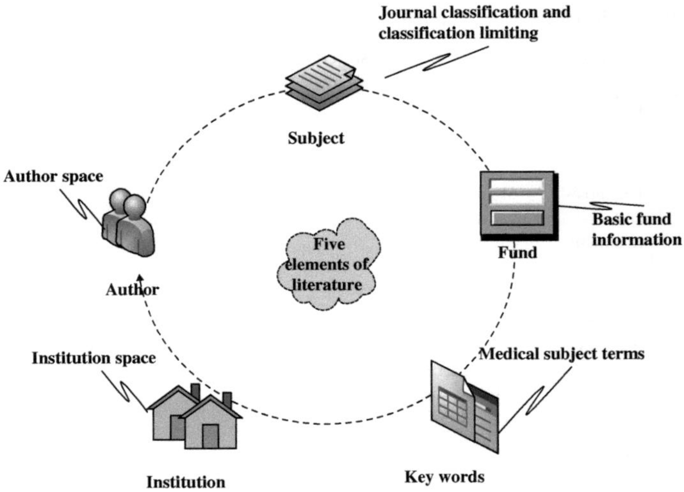

Fig. 1.2 Five-element metadata framework for bibliography resources

- **Six-element metadata framework for clinical diagnostic supporting system**: The clinical diagnostic knowledge base of Wanfang Med Online integrates six independent knowledge bases, including disease, symptoms, examination, drug, guidelines, and case report, based on ICD10 disease classification criteria. Through it, clinicians can search relevant knowledge and case reports conveniently while making clinical diagnoses. Figure 1.3 shows the information structure.

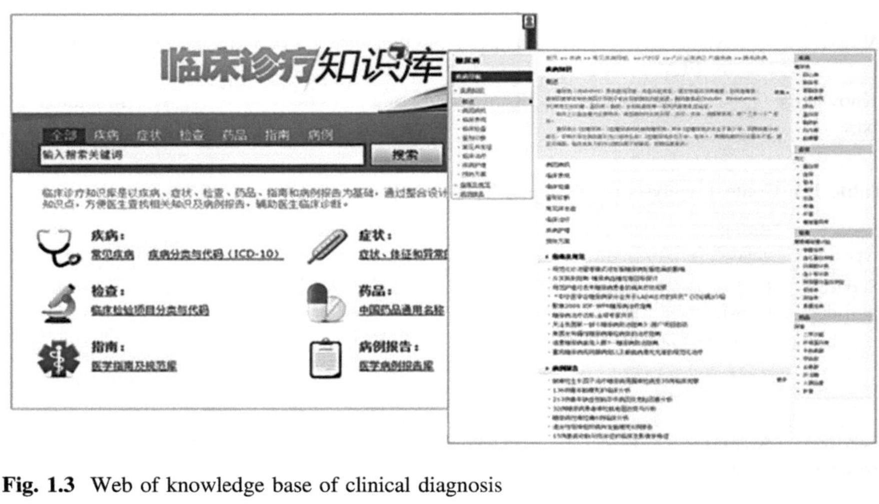

Fig. 1.3 Web of knowledge base of clinical diagnosis

Table 1.4 Classification system of Wanfang Med Online

Classification system	Target customer
Clinical department classification	Clinicians and nurses
Chinese books CLC	Librarians and journal editors
International classification of diseases (ICD10)	Clinicians, nurses, patients

Step 5. The classification system of Wanfang Med Online
A different classification system supports different point-of-information insights. Wanfang Med Online provides three main classification systems, each for a different group of users (Table 1.4).

Step 6. Application of a MeSH thesaurus
Medical Subject Headings (MeSH) is a large medical professional thesaurus that was edited and published by the U.S. National Library of Medicine (NLM) [4]; it has also been amended and supplemented regularly. The Chinese version of MeSH was translated by the Chinese Medical Institute for Scientific Information. Wanfang Med Online applied MeSH to automatic indexing and retrieval and thus improved precision and recall.

Step 7. IA Strategy for Wanfang Med Online
A website's features and style are decided by the information architecture strategy. Wanfang Med Online reflects a unique style based on proven theoretical methods, especially on content organization, structure generation, interface design, and navigation (Table 1.5).

Step 8. Project management plan for Wanfang Med Online
It seems simple to construct a website for an enterprise, but behind the website is a very complex supporting system related to the enterprise's strategy, vision, marketing system, products and service system, technical development system, and so on. Project management linked Wanfang Med Online and other elements within the company, and so determined the site's strategy and roadmap clearly (Fig. 1.4).

Step 9. Blueprint of Wanfang Med Online

Table 1.5 IA Strategy for Wanfang Med Online

Name of strategy	Theoretical methods
Organizing content	LATCH (location, alphabet, time, category, hierarchy), collecting-analyzing-group, labelling-screening-standard-indexing
Generating structure	Organizational structure metaphor, function metaphor, visual metaphors
Designing interface	HCI, scenario setting
Providing navigation	Browsing, searching, asking

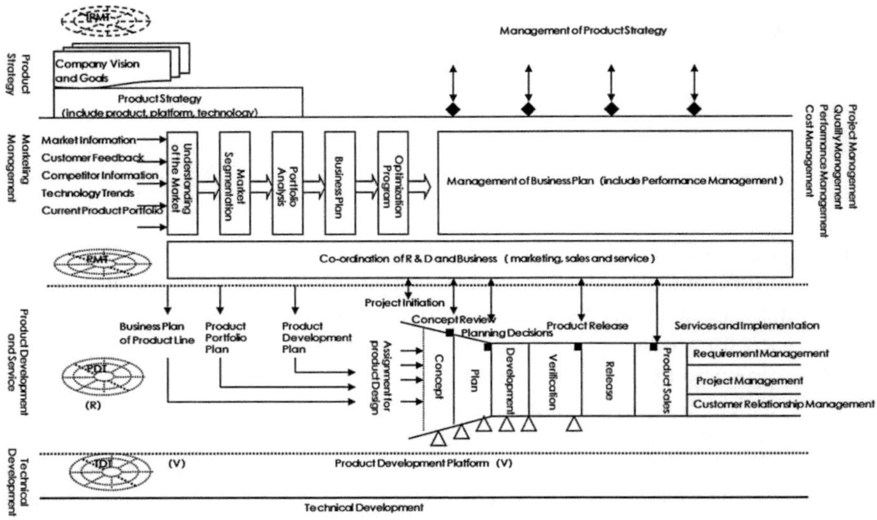

Fig. 1.4 Plan of project management for Wanfang Med Online

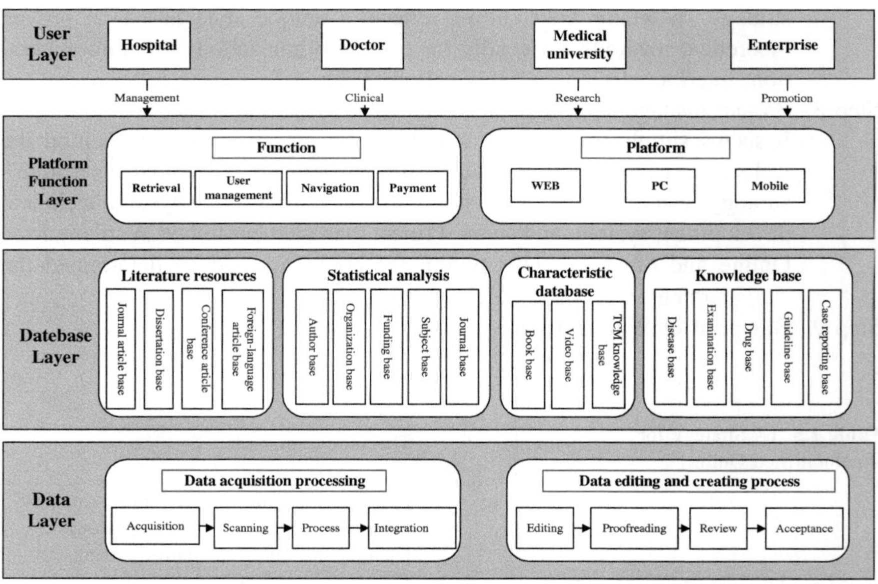

Fig. 1.5 Blueprint of Wanfang Med Online

Integrating the flood of information, multi-functions, and target users into a system in accordance with a certain logic, the website's framework and blueprint were given out as a guideline of action for all the website's stakeholders (Fig. 1.5).

Step 10. User interface

According to the users' requirements, Wanfang Med Online provides better user experience by integrating page layout and function. The features of its user interface can be summarized by three principles:

Uniformity: Unified layout and unified color. The former gives users a sense of comfort in structure, and the latter gives users color comfort.
Streamlining: Focus on the core functionality, with a premise as simple as possible to meet user demand.
Subdivision: Provide different retrieval methods according to users' various processes.

1.4 Conclusion

Based on the practice of Wanfang Med Online, the author proposed a 10-step website IA process in Chinese medical-information service, which include research on users, content analysis, context analysis, metadata framework construction, classification, thesaurus, IA strategy, PM, blueprint, and user interface, hoping to introduce as well as promote the Chinese medical-information service. It is not described in deep in many aspects and the suitability to be used in other fields still need to be further verified.

References

1. Ding W, Lin X (2009) Information architecture: the design and integration of information spaces. Morgan & Claypool Publishers, San Rafael
2. Du JT, Zhu QH, Koronios A (2014) Library and information science research in Asia-Oceania: theory and practice. IGI Global, Hershey
3. http://med.wanfangdata.com.cn
4. http://www.nlm.nih.gov/
5. Rosenfeld L, Morville P (2002) Information architecture for the world wide web, 2nd edn. O'Reilly & Associates, Inc., Sebastopol
6. Zhang XM, Cheng YH, Xu JW Assumption of construction of subject information gateway for antibacterials. J China Licensed Pharmacist 10(8):42–46
7. Zhou XY (2005) IA based on information understanding. China Renmin University Press, Beijing

Chapter 2
A Mathematical Model for Nurse Scheduling with Different Preference Ranks

Chun-Cheng Lin, Jia-Rong Kang and Wan-Yu Liu

Abstract On nurse shift schedules, it is common that the nursing staff have diverse preferences about shift rotations and days off. We propose a nurse scheduling model based upon integer programming under the constraints of the schedule, different preference ranks towards each shift and the historical data of previous schedule periods so that the nursing staff's preference satisfaction about the shift schedule is maximized. The main difference of the proposed model from the previous works is to consider that the nursing staff's satisfaction level is affected by preference ranks and their priority ordering to be scheduled, so that the quality of shift schedule is more reasonable.

Keywords Nurse shift schedules · Integer programming · Preference ranks

2.1 Introduction

This paper is concerned with the nurse schedule, which aims at determining the rotating shifts of the nursing staff over a schedule period [4]. In general, the work shifts and days off of the nursing staff should be consider in a nurse schedule. In order to ensure a feasible schedule, when the manager plans the nurse schedule, all the combinations of shifts and days off have to meet the manpower requirement of each shift, and simultaneously, the number of basic days off of each staff member should be fulfilled [3]. In addition, there are a lot of types of the work shifts and days off in the schedule. The nursing staff usually has different preferences for work

C.-C. Lin (✉) · J.-R. Kang
Department of Industrial Engineering Management, National Chiao Tung University, Hsinchu 300, Taiwan
e-mail: cclin321@nctu.edu.tw

W.-Y. Liu
Department of Tourism, Alethesia University, New Taipei City 251, Taiwan

J.J. (Jong Hyuk) Park et al. (eds.), *Future Information Technology - II*, Lecture Notes in Electrical Engineering 329,
DOI 10.1007/978-94-017-9558-6_2

shifts and days off, because diverse personal lifestyles and different degrees of physical tolerance for continuous working days.

The preference satisfaction of the nursing staff for work shifts and days off enables them to take the proper rest to increase the quality of medical service and reduce medical cost of the hospital and risks of occupational hazard [2, 5]. Therefore, it has drawn a lot of attention in the recent works to consider the nursing staff's preferences in planning the schedule of work shifts and days off, and apply maximization of satisfaction and minimization of penalty cost to evaluate the quality of shift schedule with preferences [1, 5–7].

For example, Hadwan et al. [4] investigated how to minimize the penalty cost of a nurse schedule. Aickelin and Dowsland [1] applied a genetic algorithm to solve the nurse shift schedule problem in hospitals with the aim to minimize the penalty cost for not fulfilling the preferences of the nursing staff. Maenhout and Vanhoucke [6] studied the penalty costs with multiple constraints (including the nursing staff's preferences and some certain combinations of work shifts and days off). Topaloglu and Selim [7] considered a variety of uncertain factors in nurse schedule to propose a fuzzy multi-objective integer programming model which takes into consideration the fuzziness of the objective and the nursing staff's preferences.

From the previous works, it can be found that many studies do not explore the preference ranks of the nursing staff towards each shift rotation or day off. In addition, the preferred shift and day off are not given any priority ordering. Based on this, this paper proposes a mathematical programming model for the nurse scheduling problem. Our model takes into account constraints of the schedule, different preference ranks towards each shift and the historical data of previous schedule periods to maximize the nursing staff's preference satisfaction about the shift schedule.

The main contributions of this paper are listed as follows.

- The work shift and day off preferences of the nursing staff are categorized into different levels, and are then integrated to be solved.
- A priority ordering mechanism of the nursing staff when planning their shift schedule is applied to solve the contradictory situations among their preferences.

The rest of this paper is organized as follows. Section 2.2 describes our concerned problem. Section 2.3 proposes the mathematical model. Finally, a conclusion is made in Sect. 2.4.

2.2 Problem Description

We consider a shift schedule for a 2-week work in which the shifts of a day start at midnight, and the hospital runs on a 3-shift rotation: a day shift, an evening shift, and a night shift. Note that only regular days off are planned in the schedule. The nursing staff are asked to rank their preferences for each shift and day off, which are called *preference ranks*. The preference ranks for each shift are classified into 3 types: "good," "normal," and "bad" shifts. The preference ranks of days off are classified into "good" and "bad" days off based on the "preferred" and "not preferred" days off, respectively.

In order to ensure the nurse schedule available, some schedule constraints assumed are listed as follows.

The planned schedule has some constraints on shift rotations:

- Each nursing staff member is only assigned to a fixed type of shift within each schedule period.
- The number of nursing staff members required for each shift is fixed.
- Each person should have at least an 8-h rest before continuing on to the next shift.

The planned schedule has some constraints on days off:

- The total number of days off of each nursing staff member within the schedule period is the same.
- Each nursing staff member is entitled to at least one day off each week.
- The maximum number of the nursing staff members allowed to be on day off, and the number of senior staff members working in each shift each day are known and flexible.

2.3 Methodology

In this subsection, a binary integer linear programming model is constructed, which aims to maximize the overall preference satisfaction of the nursing staff towards the shift schedule by taking into consideration the preference ranks of the nursing staff for different work shifts and days off, despite the constraints of manpower, shifts, and days off.

- Symbols

Subscript:

i	Index of a nursing staff member
j	Index of a shift type
k	Index of date

Parameters:

I	Set of the nursing staff (i.e., $i \in I$)
J	Set of shift types (i.e., $j \in J$; note that $J = \{1$ (day shift), 2 (evening shift), 3 (night shift)$\}$ in this paper)
K	Set of days off (i.e., $k \in K$; note that $\|K\| = 14$ in this paper)
M	Set of preference ranks for shifts, $M = \{1$ (good), 2 (normal), 3 (bad)$\}$

N	Set of preference ranks for days off, $N = \{1 \text{ (good)}, 3 \text{ (bad)}\}$
P	Number of considered past schedule periods
α	Coefficient of the most preferred shift, $\alpha > 1$
β	Total number of days off of each staff member within the schedule period, $\beta > 1$
R_i	The variable to identify whether staff member i is senior, $R_i = \{0 \text{ (junior)}, 1 \text{ (senior)}\}$
W_i^S	The preference weight of each nursing staff member for work shift
W_i^H	The preference weight of each nursing staff member for day off
C	The base of preference weight ($C = 2$ in this paper)
$P_{i,j}^S$	Shift preference satisfaction of staff member i in work shift j
$P_{i,k}^H$	Day off preference satisfaction of staff member i in taking day k off
$L_{i,j}^S$	Preference rank of staff member i for shift j within the schedule period, $L_{i,j}^S \in \{1(\text{good}), 2(\text{normal}), 3(\text{bad})\}$
$L_{i,k}^H$	Preference rank of staff member i for taking day k off within the schedule period, $L_{i,k}^H \in \{1(\text{good}), 3(\text{bad})\}$
$T_{i,m}^S$	In the recent P periods, the number of times staff i is assigned to the shift of preference rank m, $0 \leq T_{i,m}^S \leq P$
$T_{i,n}^H$	In the recent P periods, the number of times staff i is assigned to the day off of preference rank n, $0 \leq T_{i,n}^H \leq (P \cdot \beta)$
θ_m^S	Preference score for being assigned to the shift of preference rank m, $\theta_{m=1}^S = 1 < \theta_{m=2}^S = 2 < \theta_{m=3}^S = 3$
θ_n^H	Preference score for being assigned to the day off of preference rank n, $\theta_{n=1}^H = 1 < \theta_{n=2}^H = 3$
$D_{j,k}$	Manpower demand in shift j on day k off
$\tau_{j,k}^{\min}$	Lower bound of the required number of senior staff members in shift j on day k off
$N_{j,k}^{\max}$	The maximum number of staff members allowed to have day off in shift j on day k
$s_{i,j}^{'}$	Whether staff member i worked shift j in the previous schedule period $s_{i,j}^{'} \in \{0 \text{ (off shift)}, 1 \text{ (on shift)}\}$
$h_{i,j,k}^{'}$	Whether staff member i had day off in shift j on day k in the previous schedule period, $h_{i,j,k}^{'} \in \{0 \text{ (off shift)}, 1 \text{ (on shift)}\}$

Decision variable:

$s_{i,j}$	Whether staff i is scheduled for shift j, $s_{i,j} \in \{0 \text{ (off shift)}, 1 \text{ (on shift)}\}$.
$h_{i,j,k}$	Whether staff i is scheduled for day off j on day k, $h_{i,j,k} \in \{0 \text{ (off shift)}, 1 \text{ (on shift)}\}$

- Mathematical model

The complete mathematical model is constructed as follows.

Objective:

$$\text{Max } G = \sum_{i}^{I}\sum_{j}^{J}\left\{\left(P_{i,j}^{S}\cdot s_{i,j,}\right)+\left(\sum_{k}^{K}\left(P_{i,k}^{H}\cdot h_{i,j,k}\right)\right)\right\} \tag{2.1}$$

where

$$P_{i,j,m}^{S}=\begin{cases}\alpha W_i^S, & \text{if } L_{i,j}^S=1\\ W_i^S, & \text{if } L_{i,j}^S=2\\ 0, & \text{if } L_{i,j}^S=3\end{cases} \tag{2.2}$$

$$P_{i,k,n}^{H}=\begin{cases}\alpha W_i^H, & \text{if } L_{i,k}^H=1\\ 0, & \text{if } L_{i,k}^H=3\end{cases} \tag{2.3}$$

$$W_i^S = C^{\sum_{m=1}^{M}(T_{i,m}^{S}\cdot\theta_m^S)} \tag{2.4}$$

$$W_i^H = C^{\sum_{n=1}^{N}(\frac{T_{i,n}^{H}}{\beta}\cdot\theta_n^H)} \tag{2.5}$$

Constraints:

$$\sum_{j\in J} s_{i,j}=1,\quad \forall i\in I \tag{2.6}$$

$$\sum_{i\in I} s_{i,j}=D_{j,k},\quad \forall j\in J,k\in K \tag{2.7}$$

$$\sum_{i\in I}(R_i\cdot(s_{i,j}-h_{i,j,k}))\geq \tau_{j,k}^{\min},\quad \forall j\in J,k\in K \tag{2.8}$$

$$\sum_{k\in K} h_{i,j,k}=s_{i,j}\cdot\beta,\quad \forall i\in I,j\in J \tag{2.9}$$

$$N_{j,k}^{\max} - \sum_{i}^{I} h_{i,j,k} \geq 0, \quad \forall j \in J, k \in K \tag{2.10}$$

$$\sum_{j}^{J} \left(\sum_{k=1}^{k=7} h_{i,j,k} \cdot \sum_{k=8}^{k=14} h_{i,j,k} \right) > 0, \quad \forall i \in I \tag{2.11}$$

$$s'_{i,3} \cdot s_{i,j} \leq h'_{i,3,14} \cdot h_{i,j,k}, \quad \forall i \in I \tag{2.12}$$

$$s_{i,j} \in \{0,1\}, \quad \forall i \in I, j \in J \tag{2.13}$$

$$h_{i,j,k} \in \{0,1\}, \quad \forall i \in I, j \in J, k \in K \tag{2.14}$$

First, the objective function of the model is explained as follows. Equation (2.1) maximizes the overall preference satisfaction of the nursing staff towards work shifts and days off. The preference satisfaction of each shift preference rank of each nursing staff is calculated by Eq. (2.2) [resp., Eq. (2.3)]. The preference satisfaction calculation for shift (resp., day off) is obtained from Eq. (2.4) [resp., Eq. (2.5)].

We continue to look at the constraints of the model. Constraint (2.6) enforces each nursing staff member to be assigned to at most one work shift within each schedule period. According to Constraints (2.7) and (2.8) respectively, manpower demand and number of senior nursing staff members of each shift each day should be met. Constraint (2.9) enforces that the total number of days off assigned to each staff member in a work shift should be the same within a certain schedule period. Constraint (2.10) enforces the maximum total number of nursing staff members allowed to have day off on each shift each day. As the number of patients usually fluctuates, the schedule planner may adjust the total number of nursing staff members allowed to be on day off according to the actual number of patient on that particular day. Constraint (2.11) enforces that each staff member has to be given at least 1 day off each week, and the interval between two different shifts must be more than 8 h. According to Constraints (2.13) and (2.14), the decision variables of work shifts and days off should be either zero or one.

2.4 Conclusion

In hospital, it is specific and significant to investigate the problem of scheduling devise types of work shifts and days off for nurse staff members preferred. This paper proposes a mathematical programming model that integrates both the shift and day off into the analysis model. Not only are the shift and day off constraints taken into consideration, but also the historical records and individual preferences. Scheduling is performed on a fair and objective basis.

References

1. Aickelin U, Dowsland KA (2004) An indirect genetic algorithm for a nurse-scheduling problem. Comput Oper Res 31(5):761–778
2. Clark PA, Leddy K, Drain M, Kaldenberg D (2007) State nursing shortages and patient satisfaction: more RNs–better patient experiences. J Nurs Care Qual 22(2):119–127
3. Felici G, Gentile C (2004) A polyhedral approach for the staff rostering problem. Manage Sci 50(3):381–393
4. Hadwan M, Ayob M, Sabar NR, Qu R (2013) A harmony search algorithm for nurse rostering problems. Inf Sci 233(6):126–140
5. M'Hallah R, Sharif AA (2013) Scheduling of nurses: a case study of a Kuwaiti health care unit. Oper Res Health Care 2(1–2):1–19
6. Maenhout B, Vanhoucke M (2013) An integrated nurse staffing and scheduling analysis for loner-term nursing staff allocation problems. Omega 40(2):485–499
7. Topaloglu S, Selim H (2010) Nurse scheduling using fuzzy modeling approach. Fuzzy Sets Syst 161(11):1543–1563

Chapter 3
A New RFID and Cellular Automata Based Genetic Sequence Converter

Meng-Hsiun Tsai, Hsin-Lung Wang, Teng-Yen Wu and Mu-Yen Chen

Abstract With the development of computer technology, the more information is obtained from biological experiments through computer analysis, and even has helped give rise to a new kind of science called bioinformatics. The commonly used tool in bioinformatics is sequence alignment. Sequence alignment is a way of comparing the sequences to identify regions of similarity that may be a consequence of functional relationships between the sequences. The genetic code is highly similar among all organisms and can be expressed in a simple table with 64 entries. In this research, we uses cellular automata (CA) theory as the research topic instead of using traditional dotplot or dynamic programming to conduct sequence alignment. The parallel computing characteristic of cellular automata makes the future expansion model tremendously decrease the massive sequence computing costs. This research modifies the originally defined rules of cellular automata in order to make it more appropriate for amino acid sequence alignment.

Keywords Bioinformatics · Cellular automata · Genetic code · Sequence alignments · Parallel computing

M.-H. Tsai
Department of Management Information Systems, Institute of Genomics and Bioinformatics, National Chung Hsing University, Taichung, Taiwan, ROC

H.-L. Wang
Department of Computer Science and Information Engineering, Hungkuang University, Taichung, Taiwan, ROC

T.-Y. Wu
Microprogram Information Co., LTD., Taichung, Taiwan, ROC

M.-Y. Chen (✉)
Department of Information Management, National Taichung University of Science and Technology, Taichung, Taiwan, ROC
e-mail: mychen@nutc.edu.tw

J.J. (Jong Hyuk) Park et al. (eds.), *Future Information Technology - II*,
Lecture Notes in Electrical Engineering 329,
DOI 10.1007/978-94-017-9558-6_3

3.1 Introduction

Sequence alignment is an important tool to find the similarity of sequences. After finding the similarity of sequences (DNA sequence or protein sequence), we can know the sequences function and purpose [3, 4, 7–10].

Biological sequences can be roughly divided into three types: DNA, RNA and Protein [1]. The process by which DNA is copied to RNA is called transcription, and that by which RNA is used to produce proteins is called translation. DNA can through transcription and translation to express the genetic information in biological. Protein structure determines its function, protein function depends on the spatial structure of the protein, protein structure depends on the protein sequence, and protein structure information implicit in protein sequences. By comparing different species from the genus in the homologous proteins, the protein can be analyzed to find the species relationship, even to find their common ancestor protein. At present the first genetic code has been deciphered, but the majority of non-coding regions of DNA function are poorly understood. Bioinformatics features are: large amount of information data, information, complex and close relationship with each other. This is why we need to make sequence alignment. However, if these sequences can be converted to some graphs [2, 6, 11, 12] so that some important functions can be automatically shown and it will save a lot of time.

Through the literature and integration, we found that most of the existing published gene sequence expression method compare in a single sequence mostly, and did not feature for comparison based on Amino acids. Therefore, this research proposed an improved method and the method can be used in Amino acid sequence.

3.2 Material and Research Method

See Fig. 3.1.

3.2.1 Input Sequence

Proteins are complex organic compounds and formed by the combination of amino acids arranged in molecules. Protein is essential composition of the organism, and each of the procedures involved in cell activity. Each protein has its own unique amino acid sequences, each composed of amino acids corresponding to the nucleotides encoding and translation, and every three nucleotides decided a particular amino acid, for example, mRNA of AUG (in the DNA for ATG) translation after M, there are four RNA nucleotides in RNA, a total of $4 \times 4 \times 4 = 64$ possible, but amino acids only have 20 types, so some combination of the results are repeatable. Therefore, the same amino acid may be caused by different arrangement

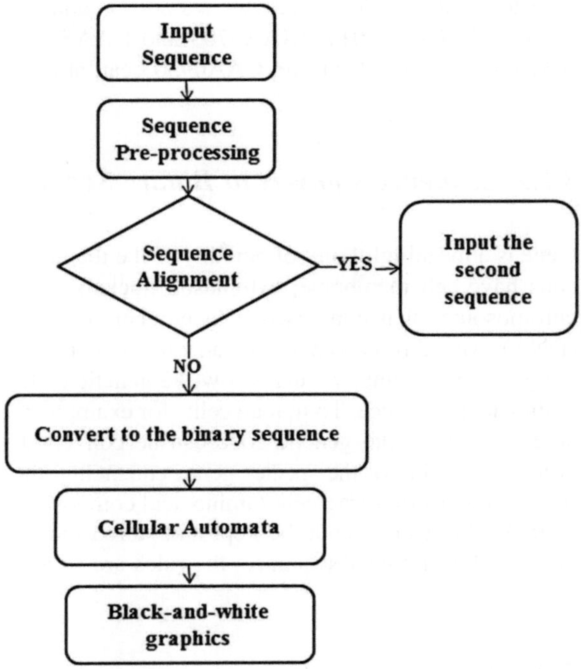

Fig. 3.1 The flow chart of the proposed method

of the nucleotides translation. In most organisms, DNA required RNA polymerase to transcription of mRNA, after transcription, mRNA can be used as a template for protein. The process from mRNA to protein called translation, the translation from the 5′ to 3′ end to find start code, you can begin to find the password after the translation, each reading until the end of the three nucleotides.

In this research, we used the input genome [5] are:

(1) AAA96714 [ATPase (*Schistosoma mansoni*)]
(2) AAC72756 [calcium ATPase 2 (*Schistosoma mansoni*)]
(3) AAD09924 [plasma membrane calcium ATPase isoform 1 (*Homo sapiens*)]
(4) AAF33531 [plasma membrane calcium ATPase isoform 3 (*Homo sapiens*)]
(5) AF361357 (Schistosoma mansoni Ca-ATPase-like protein SMA3 mRNA, complete cds)
(6) CAA09303 [calcium ATPase (*Caenorhabditis elegans*)]
(7) CAA09985 [calcium ATPase (*Caenorhabditis elegans*)]
(8) CAA59762 [calcium transporting ATPase 1 (*Saccharomyces cerevisiae*)]
(9) CAB04015 [C. elegans protein ZK256.1a, confirmed by transcript evidence (*Caenorhabditis elegans*)]
(10) KIAA0703 [Homo sapiens DNA (Celera Genomics) Homo sapiens genomic, genomic survey sequence]
(11) P20647 (Calcium-transporting ATPase sarcoplasmic reticulum type, slow twitch skeletal muscle isoform)

And AAC72756, CAA09985, P20647 and AAA96717 relatively similar; AF361357, CAB04015, KIAA0703 and CAA59762 can be attributed to a class; AAD09924, AAP33531 and CAA09303 can attributed to a class.

3.2.2 Sequence Convert to Binary Sequence

Gene is a meaningful unit of heredity. Take the human body as an example; body's cells have cell membrane, cytoplasm, nucleus and other organelles. Nucleus has chromosomes that composed of a number of protein and deoxyribonucleic acid (DNA). And chromosomes can carry the genetic code. When the species complete the gene sequencing, we can know the genetic code and it can be expressed as an amino acid sequence. To human cells, for example, the protein consists of 20 amino acids, therefore, the genetic code can be converted to the amino acid sequences. After completing of the species gene sequencing, the genetic code can be stored in the form of a string fragment (amino acid composition of the arrangement). In order to make the genetic code be kept and stored more properly, the sequence can be converted to a binary sequence, the rules are as follows in Table 3.1.

3.2.3 Implement Cellular Automata Image Process

A cellular automaton is a discrete dynamic system. Cellular automata can be re-designed in discrete time and space of macroscopic or microcosmic rules to observe the complex of space and rules. The application of cellular automata is for one-dimensional and two-dimensional currently. In this research, we used one-dimensional cellular automata. One-dimensional cellular automata consists of a set

Table 3.1 Amino acids digital coding table

Amino acid	Binary notation	Amino acid	Binary notation
P	00010	L	00100
Q	00101	H	00110
R	00111	S	01001
Y	01100	F	01011
W	01110	C	01111
T	10000	I	10010
M	10011	K	10100
N	10101	A	11001
V	11010	D	11100
E	11101	–(gap)	00000
G	11110		

of state variables according to the time change the composition S_t^i, i from 0 to N−1, N is the number of variables. Each variable can be placed in a grid, and the composition is called a cell. Each cell has different states, and in the one-dimensional cellular automata are usually have two states, 0 and 1 and the visual performance of state can be converted into black and white.

Therefore, the sequence of cellular automata array of genes can be further expressed as a barcode graphics. Among them, the sequence of the genetic code look-up table transformation method to convert into a binary sequence, through the above process by amino acids which are arranged to form the genetic code sequence will convert into a binary sequence. After the completion of converting into the binary sequence, the binary sequence will do rearrangements where the Department of sequence alignment algorithms takes the sequence of binary bits in each of three adjacent tables of values to be under the control iteration conversion. When the completion of each of three adjacent bits of the iteration is changed, then moves to the right one bit to continue iterative process, and finally generate a new sequence of a cellular automata. After completing rearrangement of the calculus sequence, select a plurality of side by side to form a binary sequence to form a selected number of rows, and then form a sequence of cellular automata array. This procedure can be simplified to the following equation:

$$\begin{aligned} D_{(i,j)} &= F\left[D_{(i-1,j-1)}, D_{(i-1,j)}, D_{(i-1,j+1)}\right], \\ &\quad \mathit{If}\ 1 \le j < S-1;\ 1 \le i < n \end{aligned} \tag{3.1}$$

$$\begin{aligned} D_{(i,0)} &= F\left[D_{(i-1,S-1)}, D_{(i-1,0)}, D_{(i-1,1)}\right], \\ &\quad \mathit{If}\ j = 0;\ 1 \le i < n \end{aligned} \tag{3.2}$$

$$\begin{aligned} D_{(i,S-1)} &= F\left[D_{(i-1,S-2)}, D_{(i-1,S-1)}, D_{(i-1,0)}\right], \\ &\quad \mathit{If}\ j = S-1;\ 1 \le i < n \end{aligned} \tag{3.3}$$

$D_{(i,j)}$ is on behalf of state value of converted the sequence of cellular automata array of (i, j) position of 0 or 1; i for the selected lines; S represents the binary sequence of length; and F refers to the rearrangement of the sequence rules of calculus functions. The sequence rearrangement calculus function F rules for the following table (Table 3.2).

Table 3.2 Iteration rule table

Rule	1	2	3	4	5	6	7	8
Before iteration	000	001	010	011	100	101	110	111
After iteration	1	0	1	1	1	0	0	0

3.2.4 Black-and-White Graphics

Showing the traditional black and white graphics, you can see with the naked eye directly after conversion of graphic sequences. We can directly observe what different proportions between the two sequences, and determine whether the two sequences are similar. It is wasting time of using the traditional method with the eye to compare the genes. Therefore, display the traditional black and white images, can determine where the difference between the two genes.

3.3 Results

3.3.1 Convert to the Binary Sequence

In the system back-end, if you input an access number, the system will determine the sequence is a protein database or a nucleotide database; if it is the protein database, the system will directly capture sequence form the NCBI database for conversion. If it is the nucleotide database, the complete data capture and search after the acquisition for the provision of translation part of the system to use. When the sequence captured, system will receive the sequence string into a binary sequences, conversion to the table in the previous section.

3.3.2 Implement Cellular Automata Image Process

When the sequence from the string into a binary number, we can begin to convert the digital sequence of images by the specified rules for cellular automata iterations. According to different rules, converted the images will not be the same. The image shows the bottom of this group under the AF361357 sequence and converted with different rules to get the images (Figs. 3.2, 3.3, 3.4, 3.5, 3.6, 3.7, 3.8 and 3.9).

Because the experimental images are a lot, there are only a few selected out of representative comparison. In this research, the section 184 rules is the most identification of all, so we chose the rules of section 184 after the experiment as a sequence alignment using the basis. Under section 184 rules, we will enter the experimental group and converted to an image sequence. With these maps, we can determine them directly from the visual similarity is very high (Figs. 3.10 and 3.11).

Figure 3.12 is the AF361357 protein fragment sequence image by CDS converted, the major conversion of the amino acid sequence capture contains 303–308 locations in the resulting image. In this fragment, it contains most of the V-shaped image on the right side, and the black lines are more obvious. Similarly, Fig. 3.13 is the CAB04015 protein fragment sequence image by CDS converted, the major conversion of the amino acid sequence capture contains 276–281 locations in the

Fig. 3.2 Image of sequence AF361357 by rule 2

Fig. 3.3 Image of sequence AF361357 by rule 5

Fig. 3.4 Image of sequence AF361357 by rule 14

Fig. 3.5 Image of sequence AF361357 by rule 18

Fig. 3.6 Image of sequence AF361357 by rule 32

Fig. 3.7 Image of sequence AF361357 by rule 43

Fig. 3.8 Image of sequence AF361357 by rule 184

Fig. 3.9 Image of sequence AF361357 by rule 247

Fig. 3.10 Image of sequence AF361357 by rule 184

Fig. 3.11 Image of sequence CAB04015 by rule 184

Fig. 3.12 Fragment image of AF361357

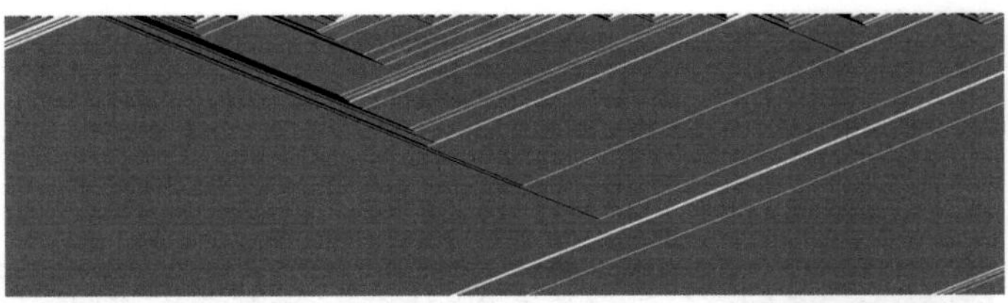

Fig. 3.13 Fragment image of CAB04015

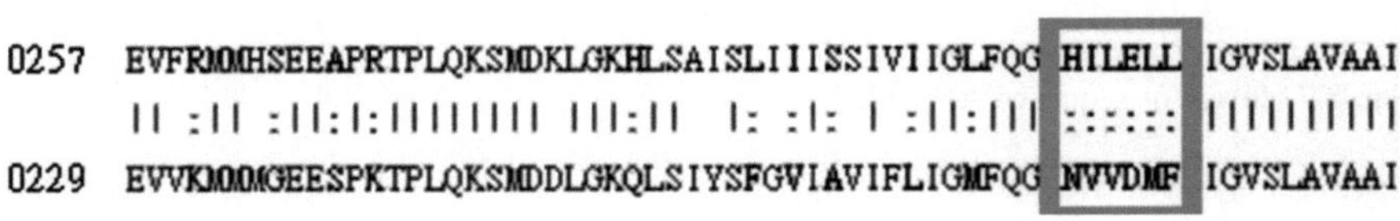

Fig. 3.14 AF361357 (*up*) and CAB04015 (*down*) different in the picture

resulting image. In this fragment, it contains most of the V-shaped image on the left side, and the white lines are more obvious. Both can be found than the latter (Fig. 3.14), the V-shaped of the Fig. 3.12 stackmore intensive, while the part of Fig. 3.13 stack more loosely.However the location 303–308 of AF361357 are for the HILELL, andthe location 276–281 of CAB04015 are for the NVVDMF, HILELLand NVVDMF are similar in chemical nature from the BLOSUM 62table and more positive points in that, so the image are generated out of the V-shaped, but in different positions.

3.4 Conclusion

Sequence alignment is a very basic tool of analyzing biological information. In this research, we used graphics conversion to replace traditional sequence alignment such as Dot plot and dynamic programming. And we compared the different amino acid sequences which were converted to graphics. By this conversion, we can easily compare the similarity between the more complex sequences. The shortcoming of traditional dynamic programming is that the results can only be known from scores. But the implement system in this research can visually identify which sequences are more similar.

Acknowledgments The authors would like to thank the reviewers for their valuable suggestions and comments that are helpful to improve the content and quality for this paper. This paper is supported by the Taichung Veterans General Hospital/National Chung Hsing University Joint Research Program, under the contract of TCVGH-NCHU1017613 and TCVGH-NCHU1027619 and the National Science Council of Taiwan, ROC, under the contract of NSC 100-2221-E-005-089- and NSC 101-2221-E-005-093-.

References

1. Adams MD, Kelley JM, Gocayne JD, Dubnick M, Polymeropoulos MH, Xiao H, Merril CR, Wu A, Olde B, Moreno RF et al (1991) Complementary DNA sequencing: expressed sequence tags and human genome project. Science 252(5013):1651–1656
2. Alston M, Johnson CG, Robinson G (2003) Colour merging for the visualization of biomolecular sequence data. In: Seventh International Conference on Information Visualization (IV 2003), pp 169–175
3. Altschul SF, Erickson BW (1986) Optimal sequence alignment using affine gap costs. Bull Math Biol 48(5–6):603–616
4. Church KW, Helfman JI (1993) Dotplot: a program for exploring self-similarity in millions of lines of text and code. J Comput Graph Stat 2:153–174
5. Da'dara A, Tsai MH, Tao LF, Marx KA, Shoemaker CB, Harn DA, Skelly PJ (2001) Schistosoma mansoni: molecular characterization of a tegumental Ca-ATPase (SMA3). Exp Parasitol 98(4):215–222
6. Hamori E, Ruskin J (1983) H curves, a novel method of representation of nucleotide series especially suited for long DNA sequences. J Biol Chem 258(2):1318–1327
7. Needleman SB, Wunsch CD (1970) A general method applicable to the search for similarities in the amino acid sequence of two proteins. J Mol Biol 48(3):443–453
8. Shibayama G, Imai H (1993) Finding K-best alignments of multiple sequences. In: Proceedings of genome informatics workshop IV, pp 120–129
9. Smith TF, Waterman MS (1981) Identification of common molecular subsequences. J Mol Biol 147(1):195–197
10. Stevens R, Goble C, Baker P, Brass A (2001) A classification of tasks in bioinformatics. Bioinformatics 17(2):180–188
11. Wu D, Roberge J, Cork DJ, Nguyen BG, Grace T (1993) Computer visualization of long genomic sequences. In: IEEE Conference on Visualization (Visualization '93), pp 308–315
12. Xiao X, Shao S, Ding Y, Huang Z, Chen X, Chou KC (2005) Using cellular automata to generate image representation for biological sequences. Amino Acids 28(1):29–35

Chapter 4
A Research on 3D Motion Database Management and Query System Based on Kinect

Edgar Chia-Han Lin

Abstract Due to the development of computer technology and the mature development of 3D motion capture technology, the applications of 3D motion databases become more and more important. How to analysis the huge data stored in the database and efficiently retrieved the matched data is an important research issue. 3D animation design is one of the important applications of 3D motion databases. Based on our teaching experience, the bottleneck of the students' learning of 3D animation is the motion animation of the 3D characters. Therefore, the 3D motion database can be used to assist the design of the motion for 3D characters. However, it is still a difficult problem because of the high complexity of the matching mechanism and the difficult of user interface design. Kinect, which is developed by Microsoft, is used as a remote controller of Xbox 360 games. Because of the capability of capturing user motions, Kinect is used in this project as the user interface. The captured data can be used as the user query and the further comparison will be performed to find the matched motion data.

Keywords 3D motion database · Index structure · Query processing · Kinect

4.1 Introduction

Due to the great improvement of technology and the mature development of 3D motion capture technology, the applications of 3D motion database become more important in recent years. How to analyze the large amount of 3D motions recorded in the databases to efficiently retrieve the desired motions become an important

---

This work was partially supported by Asia University (Project No. 102-asia-50).

E.C.-H. Lin (✉)
Department of Information Communication, Asia University, Taichung, Taiwan
e-mail: edgarlin@asia.edu.tw

J.J. (Jong Hyuk) Park et al. (eds.), *Future Information Technology - II*,
Lecture Notes in Electrical Engineering 329,
DOI 10.1007/978-94-017-9558-6_4

research issue. 3D motion data is a time sequence data which is formed by the series of reference coordinate values in different locations on the human body. Due to the large amount of reference coordinate values and the characteristics of time series data, the analysis and retrieval of 3D motion data is time consuming. There are many researches which focus on the complexity reduction of the motion data or the index structure construction for the motion data to enhance the efficiency of data comparison.

In [14], the complex multidimensional time series data is separated into several smaller segments to reduce the complexity of data. Furthermore, the segments are used to reduce the difficulty of the matching mechanism. In these years, many researches focus on the representation of the motion data. In [15], the geometric relationships between reference points on the body are defined as features which are used to represent different body poses. In [13], the piecewise-linear model is used to classify different motion types. Moreover, the index structure is constructed based on the linear models. In [3, 4], the features of the motion poses will be extracted for each motion frame which map to a multi-dimension vector. The motion content is represented by those vectors while the index structure is also constructed. In [6, 12], the motion poses are represented by a hierarchical structure and the key frames are extracted to find the motion information. However, the extraction of key frames is time consuming. In [17, 19, 20], various motion database representation mechanisms are proposed.

In [5], a method to find the similar motions is proposed. Based on the index structure, the partial queries, i.e., the motions of some specific body parts, can be processed. Five index structures are constructed for the motions performed by different body parts. Moreover, a hierarchical structure of the human body is used to integrate the index structures such that the partial motion queries can be processed. A new similarity measurement is proposed in [7] to enhance the efficiency and effective of similar matching mechanism.

Since the motion data can be represented as multiple attribute (multiple reference points) time series data, many researches focus on the similarity search of multiple attributes databases. In [18], the Dynamic Time Warping (DTW) and Longest Common Subsequence (LCSS) approaches are extended as the similarity measurement of the multiple attribute data. In [1], a pivot-based index structure for combination of feature vectors is proposed. The query processing problem is transformed to a searching problem of multiple attributes data set. In [16], an aggregate nearest-neighbors retrieval algorithm is proposed for multi-points query problem. iDistance [21] is an index structure based on distance. In [2], a preprocessing procedure is proposed to construct the neighbor graph for motion database such that the nearest neighbor search can be efficiently performed. However, the preprocessing procedure is time consuming.

In our previous works [8–11], the index structures and corresponding query processing mechanisms for multiple attribute time sequence are proposed for exact or approximate matching problem.

This paper proposes a 3D motion data management query system based on Kinect. The 3D motion data will be represented as multiple multi-attribute strings and the corresponding index structure are proposed to efficiently find the matched motions. Moreover, the Kinect device is used to capture the user queries and represented as a query string for further query processing mechanism.

4.2 Data Model of Motion Data

In this section, the data model of 3D motion is proposed to represents the content of 3D motions in the database. There are two different data models which are proposed to represent the semantic and geometric meanings of 3D motions. The semantic model is constructed based on the content information of the 3D motions while the geometric data model is constructed based on the coordinates of the reference points in 3D spaces.

To construct the semantic model, the hierarchy of human body will be defined first. In the proposed semantic model, the semantic meanings of human body are defined as the following different types: Whole Body, Half Body, Hand and Foot. The type of motions corresponding to each type of human body will be further defined as: Stand, Squat, Walk, Run, Jump, Wave and Kick. Each motion in the database will be analyzed and hierarchically classified into some particular semantic class such as Hand → Wave. Therefore, each motion in the database will be structurally annotated based on the semantic data model. The annotated metadata can be used as the query criteria specified by users.

The geometric data model is constructed based on the motion properties of the reference points in the 3D spaces. The movement of each reference point can be represented by the feature values of 3 dimensions. That is, the feature values are used to represent the movement characteristic of three different dimensions. The feature values used to represent the motion data are shown as follows.

X-dimension	**L**eft	**S**teady	**R**ight
Y-dimension	**U**p	**S**teady	**D**own
Z-dimension	**F**orward	**S**teady	**B**ackward

For example, the movement of the reference point corresponding to left wrist in Fig. 4.1 can be represented as the following 3-attribute string.

L	R	R
U	D	D
F	F	B

Therefore, the 3D motions can be represented as a set of 3-attribute strings which can be used to construct the index structure for further query processing mechanism.

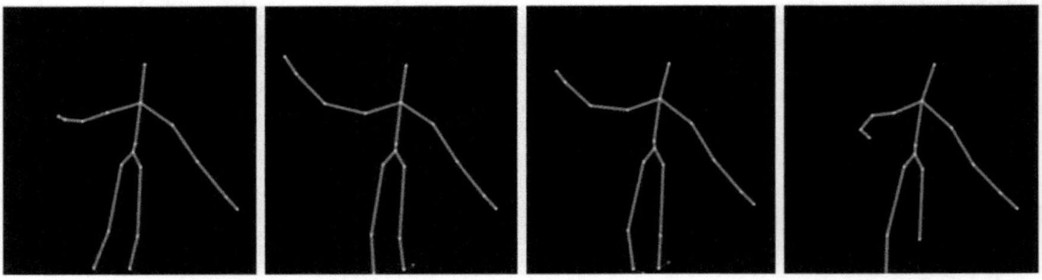

Fig. 4.1 Example of a 3D motion

4.3 Index Structure and Query Processing

Since the 3D motion data is represented as a multiple 3-attribute strings, the suffix tree based index structure proposed in our previous work [10] can be modified and applied to efficiently find the matched motion data. On the other hand, the semantic hierarchy can be used to record the semantic meaning of each motion which can be used to find the matched results by traverse the hierarchy. The architecture of the index structure construction can be shown in Fig. 4.2.

A Kinect based query interface is also designed for user to specify queries. The feature values of the query motion will be extracted and represented as a query string. The query processing mechanism will be applied to process the query string

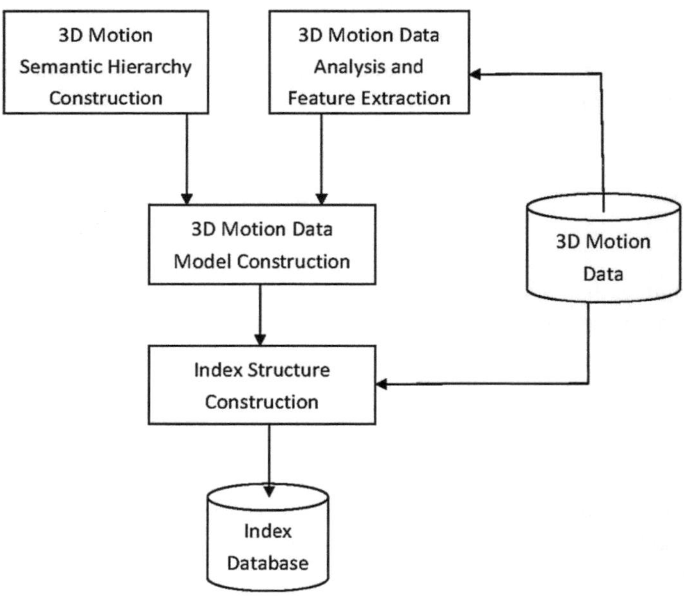

Fig. 4.2 Index structure construction

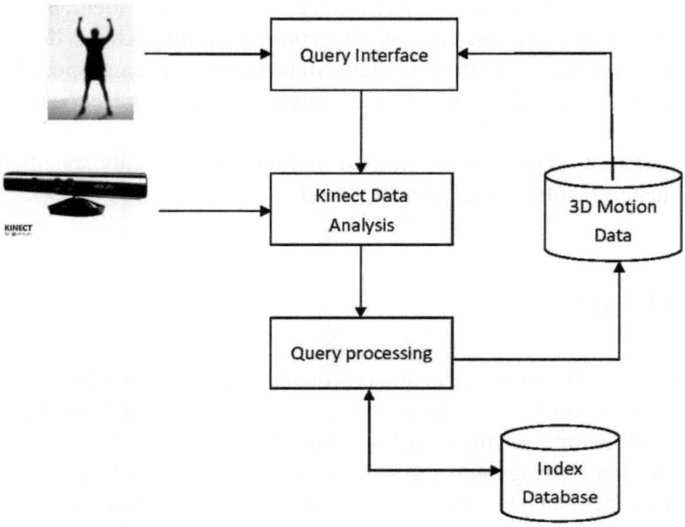

Fig. 4.3 The query processing mechanism

and the match results will be efficiently found via the index structure. The architecture of the proposed query processing mechanism is shown in Fig. 4.3.

The semantic query can be specified by users via the interface to find the corresponding motions. The geometric queries will be specified by using the Kinect device. The query motion can be acted by user and the Kinect device will capture the user motion and transform into the multiple 3-attribute strings as the query string. The multiple 3-attribute strings will be decomposed and traverse the corresponding index tree to find the matched data. At last, the matched data will be compared and the matched results will be found. Since only the movement of each reference point is considered, the absolute location of the human body captured by Kinect device won't affect the matching results. The experiment results show the flexibility of specifying queries.

4.4 Experimental Results

In this paper, a 3D data management query system is proposed. The desired motions can be specified semantically or geometrically by users. A semi-automatically annotation system is developed to annotate the semantic metadata of each 3D motion recorded in the database. Moreover, the multiple 3-attribute strings used to represent each 3D motion are automatically generated by considering the relationship of the coordinates between reference points. Then, the multiple index structures corresponding to multiple reference points are constructed according the corresponding 3-attribute string. The query interface is designed for user to specified desired motions as queries. The semantic queries can be manually specified via the interface. Moreover, the geometric queries can be captured via the Kinect device and the corresponding multiple 3-attribute strings will be transformed and specified as query strings. Therefore, the index structures are traversed and the matched results can be found.

The experiments show that the desired motions can be easily specified by users and the matched results can efficiently found.

4.5 Conclusion

In this paper, a 3D motion data management query system based on Kinect is proposed. The semantic meaning of 3D motions will be hierarchically annotated. Moreover, the geometric meanings of the 3D motions will be represented as multiple 3-attribute strings and the corresponding index structure are proposed to efficiently find the matched motions. Moreover, the Kinect device is used to capture the user queries and represented as a query string for further query processing mechanism. The experimental results show that the desired motions can be easily specified and the matched results can be found efficiently. Since the user queries may not exactly describe the desired motions, the approximate results should be further considered. We are currently working on extending the proposed methodology to find the approximate results. The similarity measurement and the corresponding matching algorithm are currently under development.

References

1. Bustos B, Keim D, Schreck T (2005) A pivot-based index structure for combination of feature vectors. In: Proceedings of SAC 2005, New York, pp 1180–1184
2. Chai J, Hodgins JK (2005) Performance animation from low-dimensional control signals. ACM Trans Graph 24(3):686–696 (SIGGRAPH 2005)
3. Chao S-P, Chiu C-Y, Chao J-H, Ruan Y-C, Yang S-N (2003) Motion retrieval and synthesis based on posture features indexing. In: Proceedings of 5th international conference computational intelligence and multimedia applications, Sept 2003, pp 266–271
4. Chiu C-Y, Chao S-P, Wu M-Y, Yang S-N, Lin H-C (2004) Content based retrieval for human motion data. J Vis Commun Image Represent 15:446–466
5. Gaurav NP, Balakrishnan P (2009) Indexing 3-D human motion repositories for content-based retreival. IEEE Trans Inf Technol Biomed 13(5):802–809
6. Gu Q, Peng J, Deng Z (2009) Compression of human motion capture data using motion pattern indexing. Comput Graph Forum 28(1):1–12
7. Kruger B, Tautges J, Weber A, Zinke A (2010) Fast local and global similarity searches in large motion capture databases. In: Proceedings of 2010 ACM SIGGRAPH/Eurographics symposium on computer animation
8. Lin EC-H (2013) Research on sequence query processing techniques over data streams. Appl Mech Mater 284–287:3507–3511

9. Lin EC-H (2013) Research on multi-attribute sequence query processing techniques over data streams. In: 2nd international conference on advanced computer science applications and technologies
10. Lin C-H, Chen ALP (2006) Indexing and matching multiple-attribute strings for efficient multimedia query processing. IEEE Trans Multimedia 8(2):408–411
11. Lin C-H, Chen ALP (2006) Approximate video search based on spatio-temporal information of video objects. In: The first IEEE international workshop on multimedia databases and data management, pp 13–20
12. Liu F, Zhuang Y, Wu F, Pan Y (2003) 3D motion retrieval with motion index tree. Comput Vis Image Underst 92:265–284
13. Liu G, Zhang J, Wang W, McMillan L (2005) A system for analyzing and indexing human-motion databases. In: Proceedings of 2005 ACM SIGMOD international conference management of data, pp 924–926
14. Lu C, Ferrier NJ (2003) Automated analysis of repetitive joint motion. IEEE Trans Inf Technol Biomed 7(4):263–273
15. Muller M, Roder T, Clausen M (2005) Efficient content-based retrieval of motion capture data. ACM Trans Graphic (TOG) 24:667–685
16. Papadias D, Tao Y, Mouratidis K, Hui CK (2005) Aggregate nearest neighbor queries in spatial databases. ACM Trans Database Syst 30(2):529–576
17. Tang K, Leung H, Komura T, Shum HPH (2008) Finding repetitive patterns in 3D human motion captured data. In: Proceedings of 2nd international conference ubiquitous information management and communication, Suwon, Korea, pp 396–403
18. Vlachos M, Hadjieleftheriou M, Gunopulos D, Keogh E (2003) Indexing multi-dimensional time-series with support for multiple distance measures. In: Proceedings of SIGMOD, Aug 2003, pp 216–225
19. Wang J, Fleet D, Hertzmann A (2008) Gaussian process dynamical models for human motion. IEEE Trans Pattern Anal Mach Intell 30(2):283–298
20. Wang X, Yu Z, Wong H (2009) 3D motion sequence retrieval based on data distribution. In: Proceedings of 2008 IEEE international conference on multimedia expo, pp 1229–1232
21. Yu C, Ooi BC, Tan K-L, Jagadish HV (2001) Indexing the distance: an efficient method to KNN processing. In: Proceedings of VLDB, San Francisco, CA, pp 421–430

Chapter 5
A SAT Approach to Nursing Scheduling Problem

Tsung-Shun Wu, Chia-Ling Ho and Chien-Liang Chen

Abstract The objective of nursing scheduling problem (NSP) is finding a shift schedule for nurses. According to the nurse's preference, the working constraints in the hospital and many possible combinations, the NSP become a complex problem and difficult to solve. The goal of NSP is to find a shift schedule in an efficiently way is very important when the problem become more complex. In this paper, we propose that the NSP can be converted into the boolean satisfiability problem (SAT). A shift schedule can be obtained by integer linear programming (ILP) method where satisfy the nurse's preference and the working constraints. We show that an example is solved by SAT approach efficiently.

Keywords Nurse shift schedule · Boolean satisfiability · Integer linear programming

5.1 Introduction

The nursing scheduling problem (NSP) is an important problem for the hospital due to the nursing shortage. How to arrange a reasonable nurse's shift schedule that satisfy all requirements is very significant. In this problem, there are many aspect that needs to be considered according to how many nurses are required in a shift, the nurse's preference and the working constraints. Hence, the NSP is a complex

---

T.-S. Wu
Department of Business Administration, Aletheia University, New Taipei City, Taiwan

C.-L. Ho
Department of Marketing and Logistics Management, Taipei Chengshih University of Science and Technology, Taipei, Taiwan

C.-L. Chen (✉)
Department of Computer Science and Information Engineering, Aletheia University, New Taipei City, Taiwan
e-mail: clchen@au.edu.tw

J.J. (Jong Hyuk) Park et al. (eds.), *Future Information Technology - II*,
Lecture Notes in Electrical Engineering 329,
DOI 10.1007/978-94-017-9558-6_5

problem and it is difficult to be solved. As we know, the complexity of NSP is a NP-Hard [7] in mathematical optimization due to enormous number of constraints and many possible solution.

The NSP has been studied in many approaches, e.g. genetic algorithm [1], local search algorithm [2], linear programming [11], integer programming [5], dynamic programing [6] and fuzzy method [10]. In this paper, we propose a boolean satisfiability (SAT) approach [9] to solve the NSP. This approach examines the whole space which is described the NSP. If NSP is unsatisfiable then it has no solution, or else it's satisfied then SAT approach will search all possible solution to find the optimal solution.

SAT problem is to determine a satisfying variable assignment for a boolean function. It is consist of a set of boolean variables and a set of constraints stated in product of sum form, i.e. the Conjunctive Normal Form (CNF), that can be solved by SAT solver. A SAT solver construct the variable assignment that satisfies CNF when it is satisfiable. An extension of SAT solver is Pseudo-Boolean (PB) constraints [4, 8] which is linear inequalities of boolean variable in the following format:

$$a_1x_1 + a_2x_2 + \cdots + a_nx_n \geq b$$

where $a_1, b \in \mathbb{Z}$ and $x_i \in \{0, 1\}$. We can translate any CNF clause to be a PB constraint. e.g., clause $(a \vee b \vee c)$ is the same as $a + b + c \geq 1$. PB constraints is the efficient way to describe NSP. In this paper, we use CNF and PB constraints to describe NSP. Then, we solve it using integer linear programming method [3].

The rest of this paper is organized as follows. Section 5.2 describes our concern problem. Section 5.3 uses a example to describe the mathematic model. Finally a conclusion is given in Sect. 5.4.

5.2 Problem Formulation

Consider a Nurse Scheduling Problem (NSP) has m nursing staffs, shifts s per day, a shift schedule for d days and needs w nursing staffs to be working in each shift. In this problem, we focus on the three shifts per day: a day shift, an evening shift and a night shift, i.e., s is three. If a nurse staff i is working, we say N_i is one, otherwise N_i is zero. If a nurse staff i is working on day j, we say $N_{i,j}$ is one, otherwise $N_{i,j}$ is zero. If a nurse staff i is working at shift s on day j, we say $N_{i,j,s}$ is one, otherwise $N_{i,j,s}$ is zero.

For NSP, two kinds of main constraints must be taken into account.

- Working constraints: It needs to make sure that w nurse staffs are assigned to each shift. It can be express in the following constraint:

$$\sum_{i=1}^{m} N_{i,j,s} = w, \ \forall j, \ \forall s \text{ and } w \leq m \tag{5.1}$$

- Nurse staff constraints: The following constraint ensures that a nurse staff is assigned to any shift if and only if the nurse staff is working.

$$\bigcup_{s=1}^{3} N_{i,j,s} \Leftrightarrow N_{i,j}, \ \forall i \text{ and } \forall j \tag{5.2}$$

- The following constraint is that a nurse staff is working during any day if and only if the nurse staff is working.

$$\bigcup_{j=1}^{d} N_{i,j} \Leftrightarrow N_{i}, \ \forall i \tag{5.3}$$

In this work, we concern the nurse staff preferences. The following constraints can be assigned to each nurse staff depending on their preferences.

- A nurse staff cannot work for two consecutive days.

$$N_{i,j} \Rightarrow \overline{N_{i,j+1}}, \ \forall i, \ \forall j \text{ and } j \neq d \tag{5.4}$$

- A nurse staff only can work up to one shift per day.

$$\sum_{s=1}^{3} N_{i,j,s} \leq 1, \ \forall i \text{ and } \forall j \tag{5.5}$$

- A nurse staff must work not less l days and no more than u days.

$$l \leq \sum_{j=1}^{d} N_{i,j} \text{ and } \sum_{j=1}^{d} N_{i,j} \leq u, \ \forall i \tag{5.6}$$

- If a nurse staff i cannot work on day j, the constraint can be expressed

$$N_{i,j} = 0 \tag{5.7}$$

- If a nurse staff i is unable to work in the day shift, it means s is one, the constraint can be expressed

$$\sum_{j=1}^{d} N_{i,j,1} = 0 \tag{5.8}$$

From above constraints, the preference of nurse staff can be considered optional and can be included when necessary. The constraints described the NSP can be seen as a set of linear constraint that can be solved by integer linear programming method.

5.3 Example

We consider an example consisting of three nurse staffs, a shift schedule for three days and three shifts per day. Each shift must be assigned to a nurse staff. Figure 5.1 shows all boolean variables in this example. Each node denotes a boolean variable. For example, N_1 and N_2 denote the nurse staff 1 and 2. $N_{2,3}$ denotes the nurse staff 2 working in day 3. $N_{1,2,3}$ denotes the nurse staff 1 working in the late shift on day 2.

The working constraints show that one nurse staff is assigned to each shift in this system. The following constraints is obtained from SAT:

$$\begin{cases} N_{1,1,1} + N_{2,1,1} + N_{3,1,1} = 1 \\ N_{1,1,2} + N_{2,1,2} + N_{3,1,2} = 1 \\ N_{1,1,3} + N_{2,1,3} + N_{3,1,3} = 1 \\ N_{1,2,1} + N_{2,2,1} + N_{3,2,1} = 1 \\ N_{1,2,2} + N_{2,2,2} + N_{3,2,2} = 1 \\ N_{1,2,3} + N_{2,2,3} + N_{3,2,3} = 1 \\ N_{1,3,1} + N_{2,3,1} + N_{3,3,1} = 1 \\ N_{1,3,2} + N_{2,3,2} + N_{3,3,2} = 1 \\ N_{1,3,3} + N_{2,3,3} + N_{3,3,3} = 1 \end{cases} \tag{5.9}$$

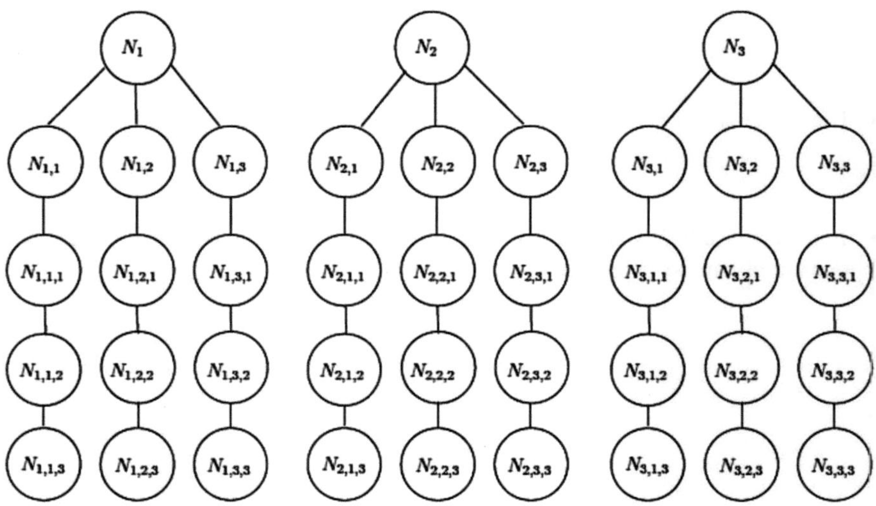

Fig. 5.1 Boolean variables in the example

The nurse staff constraints ensure that if a nurse staff is assigned to a shift then the nurse staff is working. The logic repressions and CNF clauses of nurse staff constraints can be expressed in the following:

$$\begin{array}{ll}
\overbrace{\qquad\qquad\qquad\qquad}^{\text{logic expression}} & \overbrace{\qquad\qquad\qquad\qquad}^{\text{CNF clause}} \\
N_{1,1,1} \vee N_{1,1,2} \vee N_{1,1,3} \Rightarrow N_{1,1} & (\overline{N_{1,1,1}} \vee N_{1,1})(\overline{N_{1,1,2}} \vee N_{1,1})(\overline{N_{1,1,3}} \vee N_{1,1}) \\
N_{1,2,1} \vee N_{1,2,2} \vee N_{1,2,3} \Rightarrow N_{1,2} & (\overline{N_{1,2,1}} \vee N_{1,2})(\overline{N_{1,2,2}} \vee N_{1,2})(\overline{N_{1,2,3}} \vee N_{1,2}) \\
N_{1,3,1} \vee N_{1,3,2} \vee N_{1,3,3} \Rightarrow N_{1,3} & (\overline{N_{1,3,1}} \vee N_{1,3})(\overline{N_{1,3,2}} \vee N_{1,3})(\overline{N_{1,3,3}} \vee N_{1,3}) \\
N_{2,1,1} \vee N_{2,1,2} \vee N_{2,1,3} \Rightarrow N_{2,1} & (\overline{N_{2,1,1}} \vee N_{2,1})(\overline{N_{2,1,2}} \vee N_{2,1})(\overline{N_{2,1,3}} \vee N_{2,1}) \\
N_{2,2,1} \vee N_{2,2,2} \vee N_{2,2,3} \Rightarrow N_{2,2} & (\overline{N_{2,2,1}} \vee N_{2,2})(\overline{N_{2,2,2}} \vee N_{2,2})(\overline{N_{2,2,3}} \vee N_{2,2}) \\
N_{2,3,1} \vee N_{2,3,2} \vee N_{2,3,3} \Rightarrow N_{2,3} & (\overline{N_{2,3,1}} \vee N_{2,3})(\overline{N_{2,3,2}} \vee N_{2,3})(\overline{N_{2,3,3}} \vee N_{2,3}) \\
N_{3,1,1} \vee N_{3,1,2} \vee N_{3,1,3} \Rightarrow N_{3,1} & (\overline{N_{3,1,1}} \vee N_{3,1})(\overline{N_{3,1,2}} \vee N_{3,1})(\overline{N_{3,1,3}} \vee N_{3,1}) \\
N_{3,2,1} \vee N_{3,2,2} \vee N_{3,2,3} \Rightarrow N_{3,2} & (\overline{N_{3,2,1}} \vee N_{3,2})(\overline{N_{3,2,2}} \vee N_{3,2})(\overline{N_{3,2,3}} \vee N_{3,2}) \\
N_{3,3,1} \vee N_{3,3,2} \vee N_{3,3,3} \Rightarrow N_{3,3} & (\overline{N_{3,3,1}} \vee N_{3,3})(\overline{N_{3,3,2}} \vee N_{3,3})(\overline{N_{3,3,3}} \vee N_{3,3})
\end{array} \tag{5.10}$$

The only if part of the nurse staff constraints is expressed in the following:

$$\begin{array}{ll}
\overbrace{\qquad\qquad\qquad\qquad}^{\text{logic expression}} & \overbrace{\qquad\qquad\qquad\qquad}^{\text{CNF clause}} \\
N_{1,1} \Rightarrow N_{1,1,1} \vee N_{1,1,2} \vee N_{1,1,3} & (\overline{N_{1,1}} \vee N_{1,1,1} \vee N_{1,1,2} \vee N_{1,1,3}) \\
N_{1,2} \Rightarrow N_{1,2,1} \vee N_{1,2,2} \vee N_{1,2,3} & (\overline{N_{1,2}} \vee N_{1,2,1} \vee N_{1,2,2} \vee N_{1,2,3}) \\
N_{1,3} \Rightarrow N_{1,3,1} \vee N_{1,3,2} \vee N_{1,3,3} & (\overline{N_{1,3}} \vee N_{1,3,1} \vee N_{1,3,2} \vee N_{1,3,3}) \\
N_{2,1} \Rightarrow N_{2,1,1} \vee N_{2,1,2} \vee N_{2,1,3} & (\overline{N_{2,1}} \vee N_{2,1,1} \vee N_{2,1,2} \vee N_{2,1,3}) \\
N_{2,2} \Rightarrow N_{2,2,1} \vee N_{2,2,2} \vee N_{2,2,3} & (\overline{N_{2,2}} \vee N_{2,2,1} \vee N_{2,2,2} \vee N_{2,2,3}) \\
N_{2,3} \Rightarrow N_{2,3,1} \vee N_{2,3,2} \vee N_{2,3,3} & (\overline{N_{2,3}} \vee N_{2,3,1} \vee N_{2,3,2} \vee N_{2,3,3}) \\
N_{3,1} \Rightarrow N_{3,1,1} \vee N_{3,1,2} \vee N_{3,1,3} & (\overline{N_{3,1}} \vee N_{3,1,1} \vee N_{3,1,2} \vee N_{3,1,3}) \\
N_{3,2} \Rightarrow N_{3,2,1} \vee N_{3,2,2} \vee N_{3,2,3} & (\overline{N_{3,2}} \vee N_{3,2,1} \vee N_{3,2,2} \vee N_{3,2,3}) \\
N_{3,3} \Rightarrow N_{3,3,1} \vee N_{3,3,2} \vee N_{3,3,3} & (\overline{N_{3,3}} \vee N_{3,3,1} \vee N_{3,3,2} \vee N_{3,3,3})
\end{array} \tag{5.11}$$

The following logic expressions and CNF clauses ensure that a nurse staff is working in a day only if the nurse staff is working. The constraints come from only if part of Eq. 5.3, it can be expressed in the following:

$$\begin{array}{ll}
N_{1,1} \vee N_{1,2} \vee N_{1,3} \Rightarrow N_1 & (\overline{N_{1,1}} \vee N_1)(\overline{N_{1,2}} \vee N_1)(\overline{N_{1,3}} \vee N_1) \\
N_{2,1} \vee N_{2,2} \vee N_{2,3} \Rightarrow N_2 & (\overline{N_{2,1}} \vee N_2)(\overline{N_{2,2}} \vee N_2)(\overline{N_{2,3}} \vee N_2) \\
N_{3,1} \vee N_{3,2} \vee N_{3,3} \Rightarrow N_3 & (\overline{N_{3,1}} \vee N_3)(\overline{N_{3,2}} \vee N_3)(\overline{N_{3,3}} \vee N_3)
\end{array} \tag{5.12}$$

The following logic expressions and CNF clauses ensure that if a nurse staff is working in a day then the nurse staff is working. The constraints come from if part of Eq. 5.3, it can be expressed in the following:

$$\begin{array}{ll}
N_1 \Rightarrow N_{1,1} \vee N_{1,2} \vee N_{1,3} & (\overline{N_1} \vee N_{1,1} \vee N_{1,2} \vee N_{1,3}) \\
N_2 \Rightarrow N_{2,1} \vee N_{2,2} \vee N_{2,3} & (\overline{N_2} \vee N_{2,1} \vee N_{2,2} \vee N_{2,3}) \\
N_3 \Rightarrow N_{3,1} \vee N_{3,2} \vee N_{3,3} & (\overline{N_3} \vee N_{3,1} \vee N_{3,2} \vee N_{3,3})
\end{array} \tag{5.13}$$

We also can express the nurse staff's preferences which can be expressed in the following:

- If the nurse staff 1 cannot work for two consecutive days. It can be expressed in the following logic expressions:

$$N_{1,1} \Rightarrow \overline{N_{1,2}} \text{ and } N_{1,2} \Rightarrow \overline{N_{1,3}}$$

  The CNF clauses can be expressed in the following:

$$(\overline{N_{1,1}} \vee \overline{N_{1,2}}) \text{ and } (\overline{N_{1,2}} \vee \overline{N_{1,3}})$$

- The nurse staff 1 works for four shifts during 3 days.

$$(N_{1,1,1} + N_{1,1,2} + N_{1,1,3}) + (N_{1,2,1} + N_{1,2,2} + N_{1,2,3}) + (N_{1,3,1} + N_{1,3,2} + N_{1,3,3}) \leq 4$$

- The nurse staff 1 cannot work in day shift at day 1 and day 3.

$$N_{1,1,1} + N_{1,3,1} = 0$$

- The nurse staff 1 cannot work at day 1.

$$N_{1,1} = 0$$

- The nurse staff 1 must work at least 2 days.

$$N_{1,1} + N_{1,2} + N_{1,3} \geq 2$$

From above constraints, we known that it's an integer linear constraints and can be solved it by integer linear programming method.

5.4 Conclusion

It is significant to investigate the nurse scheduling problem using boolean satisfiability approach. In this paper, we propose a SAT approach to model the nurse scheduling problem and solve it by using integer linear programming method. The proposed approach take advantage of pseudo-boolean constraints to find an optimal shift schedule that satisfies the working constraints and the nurse's preferences. Scheduling is demonstrated in an efficient way using SAT approach.

References

1. Aickelin U, Dowsland KA (2004) An indirect genetic algorithm for a nurse-scheduling problem. Comput Oper Res 31(5):761–778
2. Aickelin U, Li J (2007) An estimation of distribution algorithm for nurse scheduling. Ann Oper Res 155(1):289–309
3. Aloul F, Ramani A, Markov I, Sakallah K (2002) Generic ILP versus specialized 0-1 ILP: an update. In: IEEE/ACM international conference on computer aided design (ICCAD 2002), pp 450–457
4. Aloul FA, Ramani A, Sakallah KA, Markov IL (2007) Solution and optimization of systems of pseudo-boolean constraints. IEEE Trans Comput 56(10):1415–1424
5. Azaiez M, Sharif SA (2005) A 0-1 goal programming model for nurse scheduling. Comput Oper Res 32(3):491–507
6. Belin J (2007) Exact and heuristic methodologies for scheduling in hospitals: problems, formulations and algorithms. 4OR 5(2):157–160
7. Brucker P, Qu R, Burke E (2011) Personnel scheduling: models and complexity. Eur J Oper Res 210(3):467–473
8. Chai D, Kuehlmann A (2005) A fast pseudo-boolean constraint solver. IEEE Trans Comput Aided Des Integr Circuits Syst 24(3):305–317
9. Kundu S, Acharyya S (2008) A sat approach for solving the nurse scheduling problem. In: TENCON 2008—2008 IEEE region 10 conference, pp 1–6
10. Li J, Kwan RS (2003) A fuzzy genetic algorithm for driver scheduling. Eur J Oper Res 147 (2):334–344 (Elsevier Science B.V.)
11. Ozkarahan I (1989) A flexible nurse scheduling support system. Comput Methods Programs Biomed 30(23):145–153

Chapter 6
A Hybrid Energy-Aware Scheduling for Real-Time Task Synchronization

Da-Ren Chen

Abstract Due to the importance of resource allocation and energy efficiency, this paper considers minimizing energy consumptions and priority inversion in the embedded real-time systems. While dynamic voltage scaling (DVS) is known to reduce dynamic power consumption, it also causes increased blocking time of lower priority tasks and leakage energy consumption due to increased execution. We proposed a concept of reservation lock to prevent priority inversion using CPU's dormant mode and define a block-free interval in which both DVS and leakage-aware methods can be applied. In order to compute the optimal sleeping time and its duration and to meet the timing constraints, we also propose a weighted directed graph (WDG) to obtain additional task information. By traversing WDG, task information can be updated online and the scheduling decisions could be done in linear time complexity.

Keywords Real-time scheduling · Priority ceiling protocol · Priority inversion

6.1 Introduction

The power-aware real-time scheduling problem has been well studied, relatively few work address energy-efficient real-time task synchronization. Most embedded real-time applications have shared resources in the system and mutually exclusive access to shared resources. On such a system, real-time tasks can lead to priority inversion if a task is blocked by a lower priority task due to non-preemptive resource sharing. The recent related work is an extension of Priority Ceiling Protocol (PCP) [7] in frequency inheritance. Zhang and Chanson [9] proposed a dual-speed (DS) algorithm: One is for the execution of a task when it is not blocked,

D.-R. Chen (✉)
Department of Information Management, National Taichung University of Science and Technology, Taichung, Taiwan, ROC
e-mail: danny@nutc.edu.tw

J.J. (Jong Hyuk) Park et al. (eds.), *Future Information Technology - II*,
Lecture Notes in Electrical Engineering 329,
DOI 10.1007/978-94-017-9558-6_6

and the other is adopted to execute the task in the critical section when it is blocked. Jejurikar and Gupta [3, 4] computes two slowdown factor, which can be classified into static slowdown, computed offline based on task properties, and dynamic slowdown, computed using online task execution information. Chen and Kuo [2] proposed a DVS method using frequency locking concept which can be used to render energy-efficient to the existing real-time task synchronization protocols. The work which is designed with DVS capability to slowdown or speedup the blocked or blocking tasks in the critical section. These methods may receive additional priority inversion, and thus increase the difficulties of schedulability. Our goal is to propose a energy-aware task synchronization protocol, which can minimize the priority inversion and reduce the energy-consumption of the processor. The basic idea is to postpone the intention to the locking on resources invoked by lower-priority tasks, and to construct a blocking-free interval in which the tasks' speed can be reduced using DVS. Additionally, the extended execution due to DVS does not increase the priority inversion.

6.2 Task Model

This paper studies periodic real-time tasks that are independent during the runtime. Let $\mathcal{T}$ be the set of input periodic tasks, and n denotes the number of tasks. Each task τ_i is an infinite sequence of task instances, referred to as jobs and indexed in order of decreasing priorities. A three-tuple $\tau_i = \{T_i, D_i, C_i\}$ represents each task, where T_i is the period of the task, D_i is the relative deadline with $D_i = T_i$, and C_i denotes the worst-case execution time (WCET). The length of T_i is unique in order to have each task a unique priority index in the rate-monotonic (RM) scheduling. The jth invocation of task τ_i is denoted as $J_{i,j}$ whose actual start and finish times are denoted as $S(J_{i,j})$ and $F(J_{i,j})$, respectively. Notation $s_{i,j}$ denotes the available static slack time for job $J_{i,j}$. Job $J_{i,j}$ could be completed early at time $EC(i, j)$ during its WCET.

All tasks are scheduled on a single processor which supports two modes: dormant mode and active mode. When the processor is switched to the dormant mode, the power-consumption of the processor is assumed $S_{dorm} = 0$ by scaling the static power consumption [1], while the system clock and chipset retain necessary functions to support motoring and waking up processor at right time. To execute jobs, the processor has to be in the active mode with speed S_{active}. The time and power overhead required to switch the processor to the dormant mode can be neglected by treating them as a part of the overhead to turn on the processor. Let E_{sw} and t_{sw} denote the energy and time overhead, respectively, for switching from dormant mode to active mode. When processor is idle in the active mode, the processor executes NOP instruction at processor speed S_{idle} for low-power consumption. When processor is idle and the idle interval is longer than break-even time $\frac{E_{sw}}{P(S_{idle})}$, turning it to the dormant mode is worthwhile. The DVS model is similar to those in lpWDA [5] and can be abridged here.

We assume that semaphores are used for task synchronization. All tasks are assumed to be preemptive, while the access to the shared resources must be serialized. Therefore, task can be *blocked* by lower priority tasks. When a task has been granted access to a shared resource, it is said to be executing in its critical section [8]. The *k*th critical section of task τ_i is denoted as $z_{i,k}$ which is properly nested. Each task specifies the access to the shared resource types and the required WCETs. With the given information in [6], we can compute the maximum blocking time for a task. In the different resource synchronization protocol, such as PCP [7], each job might suffer from a different amount of blocking time from lower-priority task, due to access conflict. The goal of this paper is to propose an energy-aware real-time scheduling with task synchronization based on PCP, which can minimize the priority inversion while reduce the energy-consumption of the processor. We propose a data structure called weighted directed graph (WDG) which expresses possible priority inversion online. By traversing WDG, we can not only postpone the intention of locking on the resources invoked by lower-priority tasks but also construct a blocking-free interval to slowdown the tasks speed using DVS methods.

6.3 Motivating Example

Suppose that we have three jobs J_1, J_2 and J_3, and two shared data structures protected by the binary semaphores z_1 and z_2 in the system. In accordance with PCP, the sequence of events is depicted in Fig. 6.1a. A line at a lower level indicates that the corresponding job is in blocked or preempted by a higher-priority job while the processor mode is active. A line raised to a higher level denotes that the job is executing, and the absence of a line denotes that the job has not yet been initiated or has completed. A bold line at low level denotes that the processor has been switched in dormant mode. Suppose that

- At time t_0, J_3 is initiated and it then locks semaphore z_1.
- At time t_1, J_2 is initiated and preempts J_3.
- At time t_2, J_2 cannot lock z_1, and J_3 inherits the priority of job J_2 and resumes execution.
- At time t_3, J_1 preempts J_3 in the critical section of z_1 and executes its noncritical section code.
- At time t_4, J_1 attempts to enter its critical section z_1 and is blocked by J_3 due to priority ceiling, and J_3 inherits the priority of job J_1.
- At time t_5, J_3 exits its critical section z_1 and returns to its original priority. J_1 is awakened and locks z_1.

The priority inversions are $[t_2, t_3]$ and $[t_4, t_5]$.

This paper is motivated by the significant priority inversion and power consumption due to unused slack time and context switches. The objective is to minimize the priority inversion while reduces energy-consumption. When the available static slack time (unused time in the WCET schedule) or dynamic slack

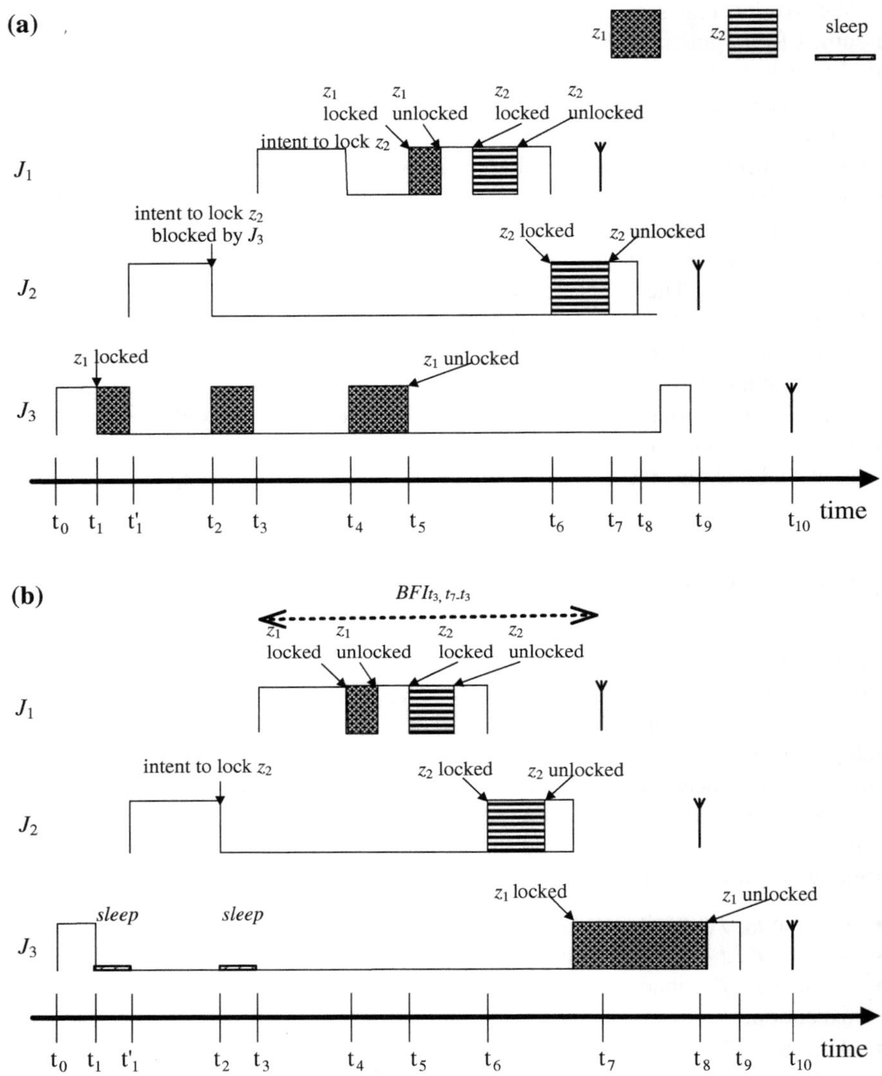

Fig. 6.1 The task synchronization of **a** primitive PCP and **b** reservation locking method

(occurred in the early-completed task) is larger than break-even time, the lower-priority task intent to lock a semaphore can be postponed until the start time of a higher-priority task. A practical approach is to postpone the task execution by switching processor to dormant mode. During the sleeping time, system still has awareness of the arrival of other jobs, and awakes processor at proper time. The example in Fig. 6.1b postpones the request of lower-priority task intent to a lock semaphore. At time t_1, J_3 has available slack in interval $[t_9, t_{10}]$ with length longer than break-even time. When a system is conscious that J_3 has intent to lock z_1, it computes the upcoming start time of higher priority tasks that might be blocked by J_3 according to PCP. In the example, J_1 and J_2 could be blocked by J_3 due to z_1, and the lengths of interval $[t_1, t_1']$ are less than the available slack. Therefore, processor switches to dormant mode at time t_1 until the start time of J_2. At time t_1', processor becomes *active* and J_2 preempts J_3 such that J_3 is still unable to lock z_1, and thus J_2 could lock z_2 at time t_2. However, J_1 could be blocked when it intends to lock z_1 if J_2 successfully lock z_2 at t_2. To further reduce priority inversion, job J_2's intent to lock z_2 should be postponed after t_3. Therefore, comparing to the result of Fig. 6.1a, the idea eliminates all priority inversions in intervals $[t_3, t_7]$.

6.4 Reservation Locking PCP

In the proposed method, PCP is extended with the concept of **reservation** locking, referred to as RL-PCP. The idea is to do pre-analysis of possible priority inversion and available slack time in the schedule. The objective is to derive the best timing and duration for switching processor to dormant mode, and thus minimize priority inversion. To understand and control the sequence of intent to lock resources, tasks are organized as a WDG reduced from the resource allocation bipartite graph in [6]. Let $G = (U, V, E)$ denote a bipartite graph whose partition of vertices has two subsets U and V. E denotes the set of edges of G, and U denote a task set $\mathcal{T}$. Let WDG($\mathcal{T}$, A) denote a weighted directed graph whose vertices in $\mathcal{T} \subseteq U$ are arranged according to their task indices. For each edge $e_{u,v} \in E$, $\tau_u \in U$ and $z_v \in V$, the set of arcs A in WDG are generated as follows

Step 1. For any pair of vertices $\tau_x, \tau_y \in U$ and $x > y$, a solid arc $a(x, y) \in A$ is directed from τ_x to τ_y if there exists two or more edges $e_{x,v}$ and $e_{y,v}$ in G where $z_v \in V$.

Step 2. For any pair of vertices $\tau_x, \tau_w \in U$ and $x > w$, a dotted arc $a(x, w) \in A$ is directed from τ_x to τ_w if there exists a vertex $\tau_y \in U$, $w > y$, and τ_x and τ_y satisfy Step 1.

Step 3. In WDG, for any pair of vertices with multiple arcs, eliminate the dotted arcs having the same blocking time as that of one of their solid arcs.

Figure 6.2 illustrates the graph reduction from bipartite to WDG. In the bipartite graph, each indirect edge is labeled with the time required to access the resources. Different from the bipartite graph, WDG has a vertex for each task but resources. A task τ_L directly blocks a higher-priority task τ_H is represented by an solid arc $a(\tau_L, \tau_H)$ from task vertex τ_L to τ_H, while an indirect block is represented by an dotted arc. In Fig. 6.2a, the bipartite graph is derived from Fig. 6.1a and can be reduced to the WDG in Fig. 6.2b. The maximum priority inversion of J_2 is an indirect blocking incurred by J_3. We may label each arc by a 3-tuple, the first element of each 3-tuple give the actual starting time of higher-priority tasks and defined as $S(\tau_H)$, while the second element gives the locking time of τ_L on semaphore z and denoted as $L(\tau_L, z)$. The last element

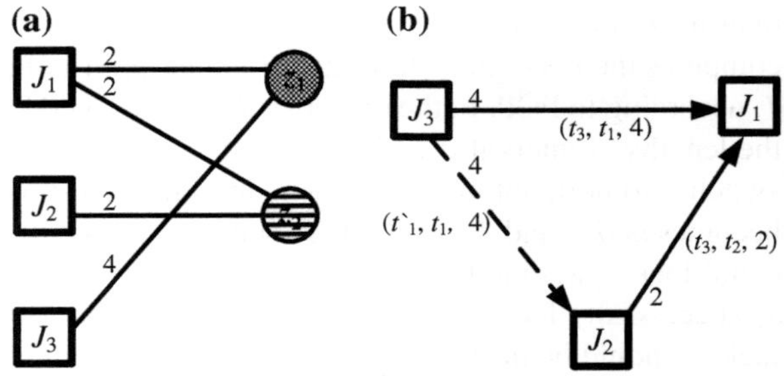

Fig. 6.2 Graph reduction from **a** bipartite graph to **b** WDG

specifies the duration of the maximum priority inversion and denoted as $I(\tau_L, \tau_H)$. The first two elements are updated during runtime while the third element is derived directly from the arcs in WDG. The example of the 3-tuple labels is illustrated in Fig. 6.2b, we have the following definitions.

Algorithm RL-PCP
Input: a set of task $\mathcal{T}$, a set of resources $\mathcal{R}$

(Offline part)
1. Reduce bipartite graph to WDG($\mathcal{T}$, A);
2. Compute the value of $I(\tau_L, \tau_H)$ with respect to each arc $a(\tau_L, \tau_H)$ in WDG;

(Online part)
On arrival of a job J_i
3. Identify a set of tasks $\mathcal{T}_H$ containing higher priority task τ_H than that of J_i;
4. Compute the value of $REW(\tau_i, \tau_H)$ for each arc $a(\tau_i, \tau_H)$ in WDG and $\tau_H \in \mathcal{T}_H$;
5. Construct a set A_i' of outgoing arcs of τ_i, $A_i' = \{a(\tau_i, \tau_H)|\alpha_{i,H} \geq \frac{E_{SW}}{P(S_{idle})}$
6. Compute the static available slack s_H for each job in $\mathcal{T}_H$;
7. Compare each s_H to the corresponding α value of the arcs in A_i';
8. Construct an arc set $A_i'' \subset A_i'$ where $A_i'' = \{a(\tau_i, \tau_x)|\ s_H \geq \alpha_{x,i},\ a(\tau_i, \tau_x) \in A_i'$ and $\tau_x \in \mathcal{T}_H\}$;
9. Search for an arc $a(\tau_i, \tau_x)$ in A_i'' with the maximum value of $REW(\tau_i, \tau_x)$ where $\tau_x \in \mathcal{T}_H$;
On beginning of one of the intervals in $ESI_{L,x}$
10. Switching the processor to S_{dorm} until the end of the interval;
On turning the processor to the active mode at time t
11. *Schedule the highest priority job in the ready queue*; On early-completion of a job at time t;
12. Compute dynamic slack time due to early completion;
On completing or beginning a job J_i at time t
13. **IF** completes early **THEN**
obtain dynamic slack s_i^d from $F(J_i) - EC(J_i)$;
14. Set $BFI = [S(J_x), F(J_x)]$ according to the recently carried out $ESI_{L,x}$;
15. Slowdown the speed of tasks whose deadlines are earlier than $F(J_x)$ using lpWDA [5];

Definition 1 In order to prevent job J_L from blocking job J_H, the expected sleep interval (*ESI*) for J_L is defined as

$$ESI_{L,H} = [L(J_L, z), S(J_H)) - \bigcup_{\forall \tau_\lambda \in \mathcal{T}, \lambda < L} NC_\lambda^{[L,H]}. \tag{6.1}$$

where $NC_{\lambda}^{[L,H]}$ denotes the set of noncritical-section intervals of job J_{λ} in interval $[L(J_L, z), S(J_H))$. The length of $ESI_{L,H}$ denotes as

$$\alpha_{L,H} = S(J_H) - L(J_L, z) - \sum_{\forall \tau\_\lambda \in I, \lambda < L} \left| NC_{\lambda}^{[L,H]} \right| \tag{6.2}$$

Definition 2 Defines the expected reduction of priority inversion (*RPI*) due to the processor sleeping in the $ESI_{L,H}$. The value of *RPI* is derived from

$$\beta_{L,H} = I(\tau_L, \tau_H) - \alpha_{L,H}. \tag{6.3}$$

According to Eqs. (6.1)–(6.3), we define a reward function for each arc in WDG.

Definition 3 A reward function for each arc in WDG is

$$REW(\tau_L, \tau_H) = \frac{\beta_{L,H}}{\alpha_{L,H}}. \tag{6.4}$$

The *reward* for an arc is referred to the reduction of priority inversion time if the processor is switched to sleep during the interval $ESI_{L,H}$. Whenever a new job J_i arrives, the value of $REW(\tau_L, \tau_i)$ with respect to each arc is refreshed. The larger the value of *REW*, the longer the priority inversion will be avoided. For example, in Fig. 6.2b, the values of $\alpha_{3,1}$ and $\alpha_{3,2}$ are set respectively $x = t_3 - t_1 - (t_2 - t_1')$ and $y = t_1' - t_1$, and the values of $\beta_{3,1}$ and $\beta_{3,2}$ are respectively $4 - x$ and $4 - y$. In accordance with Eq. 6.4, the values of *REW*(3,1) and *REW*(3,2) are $\frac{4-x}{x}$ and $\frac{4-y}{y}$, and obviously *REW*(3,1) < *REW*(3,2). Assuming that available slack for τ_3 is larger than the values of x and y, the proposed algorithm switches the processor to sleep in the duration of $[t_1, t_1']$, and traverses from vertex J_3 to J_2. In the vertex J_2, we can traverse from J_2 to J_1 by switching the processor to sleep mode in interval $[t_2, t_3]$.

Definition 4 (*Blocking-free interval, BFI*) A time interval $BFI_{t,\ell}$ is said to be blocking free in the real-time task synchronization if the interval $[t, t + \ell]$ does not have any priority inversion.

Lemma 1 *When the time at which J_L has intent to lock z is later than the $S(J_H)$, they do not give rise to priority inversion during interval $[S(J_H), F(J_H)]$.*

Proof sketch In accordance with WDG, J_H has higher priority than J_L, as soon as J_L begins after $S(J_H)$, J_L cannot lock z until J_H completes. Therefore, J_H is not preempted by J_L until the completion time of J_H. □

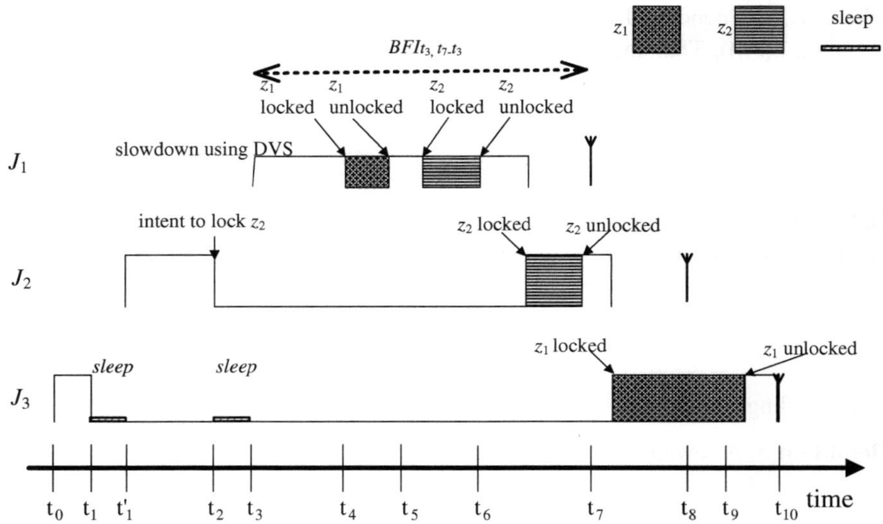

Fig. 6.3 An example using DVS scheduling

Obviously, jobs J_H and J_L do not give rise to the priority inversion in the interval $[F(J_H), D_H]$ when J_H completes before F(JH). Therefore, when the processor sleeps in interval $ESI_{L,H}$, the value of *BFI* can be derived from

$$BFI_{t,l} = [S(J_H), D_H] \tag{6.5}$$

for all jobs J_H are successors of J_L in WDG.

The purpose of BFI is to identify the jobs for saving more energy using DVS. The jobs whose deadlines are in the BFI interval can compute their available slack time to decrease their speed and satisfy their timing constraints. The slack computation such as lpWDA [5] can be applied without modification to our method. An example is presented in Fig. 6.3. By updating the information of arcs in WDG during runtime, we can traverse the WDG by following the current job and make decisions on switching the processor to active or dormant mode.

6.5 Conclusions

In the future, we will continuously improve energy-efficiency of real-time task synchronization with speed switching overhead consideration. By using DVS and leakage-aware techniques, we will decrease not only the priority inversion but also energy consumption in the real-time systems. The objective is to minimize the priority inversion and reduce both dynamic and leakage energy, provided that the schedulability of tasks is guaranteed. By traversing the vertex of WDG, the scheduling decisions can be done efficiently during the runtime. Another characteristic is that, in the proposed concept, DVS does not worsen the situation of inevitable priority inversion.

For further study, we shall explore an evaluation function that provides suggestions on how to use DVS or leakage-aware technique during runtime. Future research and experiments in these areas may benefit several mobile system designs.

References

1. Butts JA, Sohi GS (2000) A static power model for architects. In: Proceedings. of the 33rd annual international symposium on microarchitecture (MICRO-33), Monterey, California, 10–13 Dec 2000, pp 191–201
2. Chen J-J, Kuo T-W (2006) Procrastination for leakage-aware rate-monotonic scheduling on a dynamic voltage scaling processor. In: ACM SIGPLAN/SIGBED conference on languages, compilers, and tools for embedded systems (LCTES'06), Ottawa, 14–16 June 2006. ACM Press, New York, pp 153–162
3. Irani S, Shukla S, Gupta R (2003) Algorithms for power savings. In: Proceedings of the fourteenth annual ACM-SIAM symposium on discrete algorithms (DA'03), Inner Harbour, Baltimore, MD, USA, 12–14 Jan 2003. ACM Press, New York, pp 37–46
4. Jejurikar R, Gupta RK (2005) Dynamic slack reclamation with procrastination scheduling in real-time embedded systems. In: Proceedings of the 42nd design automation conference (DAC 05), San Diego, CA, USA, 13–17 June 2005. ACM Press, New York, pp 111–116
5. Kim W, Kim J, Min SL (2003) Dynamic voltage scaling algorithm for fixed-priority real-time systems using work-demand analysis. In: Proceedings of the 2003 international symposium on low power electronics and design (ISPLED'03). ACM Press, New York, pp 396–401
6. Liu JWS (2000) Real-time systems. Prentice Hall PTR, Upper Saddle River. ISBN 0130996513
7. Sha L, Rajkumar R, Lehoczky JP (1990) Priority inheritance protocols: an approach to real-time synchronization. IEEE Trans Comput 39(9):1175–1185
8. Silberschatz AP, Galvin B, Gagne G (2011) Operating system concepts. Wiley, Hoboken
9. Zhang F, Chanson ST (2002) Processor voltage scheduling for real-time tasks with non-preemptible sections. In: 23rd proceedings IEEE real-time systems symposium (RTSS 02), Austin, TX, Dec 2002. IEEE, pp 235–245

Chapter 7
Applying Taiwan EMR Exchange Architecture to Establish a Mobile-Healthcare Management System for Chronic Disease in an Aboriginal Tribe

Hsiao-Hsien Rau, Yen-Liang Lee, Chien-Yeh Hsu, Duu-Jian Tsai, Shih-Chang Chen, Li-Min Wei and Suleman Atique

Abstract The information system of Mobile-healthcare Management is using the newly and sophisticated technology to collocate physicians and nurses' professional skills. It may improve the healthcare quality of remote areas and increase their medical resources. This research aims at establishing a proper clinical medical information system. Because of most of the aboriginal tribes lack medical service, it is hard for people to visit hospital, so it need the case manager visit the patient weekly, and record the patients' situation on paper by manually, this research want to build a system to solve some problems. Consider the information device insufficiency in remote areas and need high mobility of peripatetic medical service, we use tablet PC to as a hardware and using it in a tribe which is located in the east of Taiwan. This system will make case managers get patient's basic and clinical data in their hospitals and also get these data from another hospitals through

H.-H. Rau · Y.-L. Lee · C.-Y. Hsu (✉) · D.-J. Tsai · L.-M. Wei · S. Atique
Graduate Institute of Biomedical Informatics, Taipei Medical University, Taipei, Taiwan
e-mail: cyhsu@tmu.edu.tw

H.-H. Rau
e-mail: d110097003@tmu.edu.tw

Y.-L. Lee
e-mail: lee@tmu.edu.tw

D.-J. Tsai
e-mail: dj.tsai@msa.hinet.net

L.-M. Wei
e-mail: edword@ms2.hinet.net

S. Atique
e-mail: gcufpharmd@yahoo.com

S.-C. Chen
National Council for Sustainable Development, Taipei, Taiwan
e-mail: scchen1980@ntu.edu.tw

J.J. (Jong Hyuk) Park et al. (eds.), *Future Information Technology - II*,
Lecture Notes in Electrical Engineering 329,
DOI 10.1007/978-94-017-9558-6_7

electronic medical record exchange center. Therefore, it can help case managers know the patients' conditions before they go to provide care to the patient.

7.1 Introduction

7.1.1 *Aboriginal Tribe Healthcare in Taiwan*

Taiwan has 55 aboriginal townships, including 39 mountain aboriginal townships and 25 flat aboriginal townships. There are a lot of public health problems in remote areas in Taiwan due to special environment, broad district, and different folkloric methodologies, they lead to less health care resources, lower income and unemployment. It lacks resources in the medical and health care.

For an example, Jing-Yue Tribe is a Tai-Yai Tribe situated in the mountain area in Nan-Au Village of I-Lan County. Most of the inhabitants are unemployed and make their lives by growing simple crops and hunting. Jin Yue tribe actually has about 200–300 residents, majority being elderly and children. Especially the elder residents are facing a serious problem of lack of medical service, they have only one clinic and doctor serve there for only 2 days in a week, therefore, in other days, they need to visit the hospital by car/bus and have to travel by a hilly area through very long distance. So, when they need to visit the doctor regularly to treat the disease or have medical examination, but they give up that because of inconvenient traffic or they need their children to accompany, being the major reasons which don't let these elderly residents don't want to visit physicians.

7.1.2 *Electronic Medical Record Exchange in Taiwan*

Taiwan has built a national electronic medical record exchange center (EEC) in 2011, and uses it in exchanging medical images and reports, Laboratory Examination Report, medication history, discharge summary. Besides that, EEC also can be extended to exchange others medical information, including operation record, clinical physician order and laboratory examination report etc. Hospitals upload the index of medical image and all kinds of reports to the index server which is setup in Bureau of national health Insurance, BNHI will collect and batches synchronize to EEC. Hospitals can use the integrated exchange system or login to the website of exchange center to query the index of patient's hospital visit during these 6 months. If hospital needs to take the complete medical record or download the medical image, they also can get patient's medical record from EEC under the premise of patient consent.

Until now, EEC not only effectively integrates the medical record data on a platform, but also lets the format of electronic medical record standardized, it reaches patient-centered medical data collection and lets physicians understand patients' situation in better way.

7.2 Methods

7.2.1 System Design

This system is designed in a client-server architecture, as in Fig. 7.1, Hospital (show as SMH in the figure) have their own electronic medical record, case-managers can get patient's basic data, drug use history and medication use history from their hospital information system (HIS), or electronic medical record (EMR) data warehouse, besides that, if hospital have prior authorization by the patient, they also can through electronic medical record exchange center (EEC) to get the medical record from another hospitals.

Case-managers can got the medical record of each patients before they visit the resident, system can fill many fields automatically, because the medical records had downloaded already, case-managers can use this system fill the homecare report, during this process case-manager just need few fields include patient's current situation, it let they can have more time to care their patient. After case-manager finish fill the homecare report form, system will auto detect the internet connection, if it is workable, system will use internet save the case report to the personal health data repository which is located in hospital, otherwise, system will save this as a text file (.txt) and put in the hard disk.

After case-manager back to the hospital, system can synchronize with the data repository automatically, after that, case-manager can management and view or do some statistic in the Care Portal or use these information to complete other works.

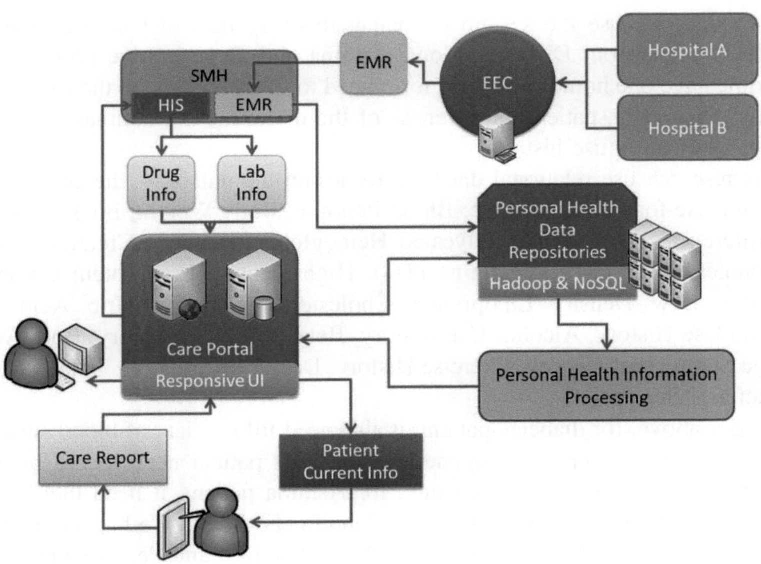

Fig. 7.1 Mobile-healthcare management system architecture

7.2.2 User Interface Design—Responsive Design Technology

We use it technology in the front end user interface design. Responsive Design is an idea of providing the user with best viewing experience of a website across devices of various sizes (from mobile phones to desktop computer monitors) [1–3], to provide an optimal viewing experience—easy reading and navigation with a minimum of resizing, panning, and scrolling [4–6, 8, 9]. It will makes the EMR/EHR web solutions' UI becomes more adaptable on multi-devices, ensures that information is displayed correctly [7].

Server-side components (RESS) in conjunction with client-side ones such as media queries can produce faster-loading sites for access over cellular networks and also deliver richer functionality/usability avoiding some of the pitfalls of device-side-only solutions [10–12].

We are using Bootstrap, Response and Lettering to archive the responsive design. Bootstrap is an open source framework for building web interfaces employing responsive design [13, 14]. Response provides tools for building performance-optimized, mobile-first responsive websites [15–17]. Lettering for controlling the appearance of your web type, is a great tool to help designers get a chokehold on their typography [18, 19].

7.2.3 Logical Architecture Design of Homecare Report Database

In this study, we use the system to management 4 kinds of Chronic Disease, it include Hypertension, Diabetes, Gout, Asthma and Degenerative arthritis, every syndrome have one homecare report form need to fill, the fields on the form include patients' basic data, patient's awareness of the disease, lab examination data, life history, medication use history.

This research use relational database to design the database, the common data fields in these four forms include: Blood Pressure, Body Weight, BMI, Abdominal Circumference, Blood Sugar, Glycated Hemoglobin (HbA1c), Creatinine (CRE), total cholesterol (TCH), Creatinine (TG), High Density Lipoprotein Cholesterol (HDL-C), Low Density Lipoprotein Cholesterol (LDL-C), Uric Acid (UA), Tobacco Use History, Alcohol Use History, Betel Nuts Use History, Body Weight Change Status, Diet History, Exercise History, Drug Use Situation, Family Support and Referral Services.

Besides above, for diabetes patient, it also need fill the data of blood sugar self-control record; for gout and Degenerative arthritis patient need record break out frequency and hospital visit frequency; for Asthma patient, it need the many lab examination results, include: Immunoglobulin E (IgE), Carbapenem-resistant Enterobacteriaceae (CRE), complete blood count (CBC) and Peak expiratory flow (PEF) etc.

7.3 Result and Discussion

7.3.1 Functions of the System

This study use tablet PC to be a hardware and using it in a aboriginal tribe. This system let case-managers can get patient's basic and clinical data in their hospitals and also get these data from another hospitals through electronic medical record exchange center. Therefore, it can help case managers know the patients' situations before they go to care the patient. Functions of the system are show as follow and case-managers need to login before they start use the system:

Show patient list: System will show name list of the patients that visited by this case-manager, system will also show the date of previous visit.
Management patients' basic data: Case-manager can click patient's ID number, system will show the basic data of this patient on the screen, if it need correct, case-manager just need to click "Revise" button and click "Save" after revise.
Fill homecare report form: Case-manager only can choice one of five kinds (hypertension, diabetes, gout, degenerative, asthma) homecare report form, case-manager can choice one to fill using drop-down list, most of fields are all design to checkbox or radio button, it let case-manager fill the form more convenience, and can reduce the probability of make mistake when fill the form.
Show or correct the previous homecare report form: Case-manager can click patient's ID number, system will show the previous homecare report form of this patient on the screen, if it need correct, case-manager just need to click "Revise" button and click "Save" after revise.
Draw a trend chart of every lab/health examination result value: System will draw trend chart automatically, case-manager can select the patient and time duration, and then system can show the chart on the screen.

7.3.2 Benefit of the System

After finishing the development of the system, we have an education workshop to target 5 case-managers and ask them use the system for 30 min; then we discuss with them, and ask them for suggestions on advantages and disadvantages of this system. The contents of discussion are recorded and reviewed by experts, and then we can conclude that this system has 7 main benefits as follow:

Legible Charting: Up-to-date, on-screen patient information will ensure consistency and eliminate handwritten data.
Comprehensive Information: Provides a secure, authorized access to a patient's complete medical history, consolidated in a single record.

Medication Management: All medications, prescribed and discontinued, are tracked using the medication management module and alerts can be created for allergic reactions or contraindications.
Highlight Essential Information: Automated reminders will highlight vital patient information such as chronic illnesses or new conditions.
Improved Quality of Care: Time saved on administration will enable case-managers to streamline processes and devote more time to patients. Immediate access to patient information means case-managers will be able to respond quickly and appropriately to emergencies.
Proactive Health Care: Access to comprehensive health data through the EMR will enable case manager can identify trends and analyze data to support decision-making and use alerts and management plans to treat patients with chronic diseases before they become ill.
Increased Privacy and Security: Secure Authentication ensures only users that are allowed to access the information are given access. User access audits logs and processes are in place to ensure the checks and measures in place for system access.
Collaborative Care: One network will allow seamless communication and referrals between nurses and other professionals in other programs.

7.3.3 Effectiveness Evaluation

After above education workshop, we let the case-manager use this system and bring tablet PC to patient's home in aboriginal tribe, every case-manager use more than 5 times and visit 27 aborigines, After statistical, before use this system, case-management use more than 30 min to get and record patients' data through viewing the medical record, and use 45 min to care the patient and fill the form, also use 45 min to generalize and arrange the forms; after use this system, these time are decrease to 13, 25, and 15 min.

References

1. On2 Technologies Inc. Advantages of TrueMotion VP6 technology. White paper, On2 technologies Feb 2004
2. Anantraman V, Mikkelsen T, Khilnani R, Kumar VS, Pentland A, Ohno-Machado Fraser H, Jazayeri D, Mitnick C, Mukherjee J, Bayona J (2002) Informatics tools to monitor progress and outcomes of patients with drug resistant tuberculosis in Peru. In: Proceedings of the AMIA annual fall symposium, pp 270–274
3. Anderson LB (2003) HHS unveils electronic plan for tracking medical records; system would allow access to patient data by doctors, hospitals, insurance firms. The Wall Street Journal, 2 July
4. Butler R (2007) The ICD-10 general equivalence mappings: bridging the translation gap from ICD-9. J AHIMA 78(9):84–86

5. Dolin RH et al (2006) HL7 clinical document architecture, Release 2. J Am Med Inform Assoc 13(1):30–39
6. Riesmeier J, Dogac A et al (2005) A survey and analysis of electronic healthcare record standards. ACM Comput Surv (CSUR) 37:277–315
7. Fraser HS, Jazayeri D, Bannach L, Szolovits P, McGrath SJ (2001) Telemedmail: free software to facilitate telemedicine in developing countries. Medinfo 10(Pt 1):815–819
8. Heitmann KU, Schweiger R, Dudeck J (2002) Discharge and referral data exchange using global standards—the SCIPHOX project in Germany. Stud Health Technol Inform 90:679–684
9. Hock C (2003) Calculating your bandwidth and software license needs for the macromedia flash communication server MX. White paper, Macromedia, June 2003
10. http://searchhealthit.techtarget.com/definition/Clinical-Document-Architecture-CDA
11. Jian W-S et al (2007) Building a portable data and information interoperability infrastructure —framework for a standard Taiwan electronic medical record template. Comput Methods Programs Biomed 88(2):102–111
12. Kunkel T (2003) Streaming media: technologies, standards, applications. Wiley, New York
13. Lynch K (2005) Macromedia flash platform. White paper, Macromedia
14. Macromedia (2003) Macromedia flash communication server MX 1.5-developing communication applications, 2nd edn. Macromedia
15. Mon DT (2003) Relational database management: what you need to know. J AHIMA 74 (6):40–45
16. Rotich JK, Hannan TJ, Smith FE et al (2003) Installing and implementing a computer-based patient record system in sub-Saharan Africa: the Mosoriot medical record system. J Am Med Inform Assoc 10:295–303
17. Spooner SA, MD, FAAPa, David C Classen, MD, MSb (2009) Data standards and improvement of quality and safety in child health care. Pediatrics vol. 123 Supplement 2 pp. S74–S79
18. Vranken R, Coulombier D, Kenyon T et al (2002) Use of a computerized tuberculosis register for automated generation of case finding, sputum conversion, and treatment outcome reports. Int J Tubercular Lung Dis 6:111–120
19. Weber R (2003) Conceptual modeling and ontology: possibilities and pitfalls. J Database Manage 14(3):1–20

Chapter 8
Automatic Classification of Digitally Modulated Signals Based on K-Nearest Neighbor

Woo-Hyun Ahn, Sun-Phil Nah and Bo-Seok Seo

Abstract In this paper, we propose an automatic classification method for eight digitally modulated signals, such as 2FSK, 4FSK, MSK, BPSK, QPSK, 8PSK, 16QAM, and 64QAM. The method uses spectral correlation density and high-order cumulants as features. For feature classification, *K*-nearest neighbor algorithm is used. Simulation results are demonstrated to evaluate the proposed scheme.

Keywords Automatic modulation classification · Cyclostationarity · K-nearest neighbor · High-order cumulant

8.1 Introduction

Automatic modulation classification (AMC) is the technique to recognize or classify the modulation format of received signal using minimum a priori knowledge of the signal. For example, AMC is used for an intermediate step between detection and demodulation of received signals in software defined radio platform. In addition, AMC has wide applications in civilian and military such as electronic warfare, surveillance, cognitive radio, and so on [2].

AMC is a complicated task because the receiver has no a priori knowledge about the received signal such as carrier frequency, signal power, and symbol rate. Furthermore, practical channel environment with multipath fading and frequency-selectivity make it more difficult. To overcome these problems, many experiments and researches are given for the last two decade.

W.-H. Ahn · B.-S. Seo (✉)
Department of Electronics Engineering, Chungbuk National University, Cheongju, Korea
e-mail: boseok@cbnu.ac.kr

W.-H. Ahn
e-mail: glingi@cbnu.ac.kr

S.-P. Nah
Agency for Defense Development of Korea, Daejeon, Korea
e-mail: spnah@add.re.kr

J.J. (Jong Hyuk) Park et al. (eds.), *Future Information Technology - II*,
Lecture Notes in Electrical Engineering 329,
DOI 10.1007/978-94-017-9558-6_8

AMC methods are normally classified into two major categories of likelihood-based approach and feature-based approach [4]. The first approach is based on the likelihood function of the received signal and uses a likelihood ratio test for the classification. Although this method shows optimal performance, the much computational complexity and high sensitivity to modeling mismatch make it hard to utilize in practical applications. On the other hand, the second approach is of low computational complexity and robust to modeling mismatch when appropriate features and classifier are designed. Also, this approach shows near optimal performance.

Generally, as features, statistical characteristics of the signal are used. In [1], the features of high-order cumulants (HOCs) with K-nearest neighbor (KNN) are proposed for classification of BPSK, QPSK, 16QAM, and 64QAM signals. It shows better performance than the AMC methods using a support vector machine or a naive-Bayes. In [6], an AMC method using spectral correlation density of cyclostationary signals is proposed. A neural network is used for classification of low-order digitally modulated signals such as AM, BSPK, CDMA, and 2FSK.

In this paper, we propose an AMC scheme to classify eight digitally modulated signals of 2FSK, 4FSK, MSK, BPSK, QPSK, 8PSK, 16QAM, and 64QAM. The scheme is based on the features of HOCs and spectral correlation density with a simple KNN classifier. The fourth-order and sixth-order cumulants are used as features. The normalized spectral correlation density is also used to characterize the cyclostationarity of the modulated signals. The features are estimated from the oversampled baseband signal.

8.2 Signal Model

At the receiver, the received signal is down converted to baseband and low-pass filtered to reject the out-of-band noise. Assuming perfect frequency and timing synchronization, the complex baseband signal can be represented as [4]

$$y(t) = A\sum_{l=-\infty}^{\infty} x(l)p(t-lT) + w(t) \tag{8.1}$$

where A represent the amplitude affected by the channel, and T is symbol duration. The l-th transmitted symbol $x(l)$ is assumed to be independent and identically distributed. The shape of the received symbol $p(t)$ reflects the channel effect, and $w(t)$ is the zero-mean additive white Gaussian noise (AWGN). We assume that the channel has been equalized and the symbol duration is assumed to be known.

8.3 Features for Modulation Classification

8.3.1 Normalized Spectral Correlation Density

If the received signal $y(t)$ is a second-order cyclostationary random process, its autocorrelation function $R_y\{t, \tau\} = E\{y(t + \tau/2)y^*(t - \tau/2)\}$ is not a function of the time lag τ but a periodic function of t. Therefore, it can be represented as a Fourier series. The Fourier transform of the Fourier series coefficients $R_y^\alpha(\tau)$, called spectral correlation density (SCD), is given by [5]

$$S_y^\alpha(f) = \int_{-\infty}^{\infty} R_y^\alpha(\tau)\mathrm{e}^{-j2\pi f\tau} d\tau \tag{8.2}$$

The SCD means a cross-correlation function between the spectral components of $y(t)$ at frequencies $f + \alpha/2$ and $f - \alpha/2$. If we assume that the received signal is cycloergodic, the normalized SCD is given by

$$C_y^\alpha(f) = \frac{S_y^\alpha(f)}{[S_y(f + \alpha/2)^* S_y(f - \alpha/2)]^{1/2}} \tag{8.3}$$

As a feature of the modulation signal, we can utilize the normalized SCD at $f = 0$. In what follows, we call $C_Y^\alpha(0)$ as the normalized SCD for convenience.

Figure 8.1 shows the normalized SCD for various modulation signals with signal to noise ratio (SNR) of 20 dB. For PSK and QAM signals, a raised-cosine filter with roll-off factor of 0.35 is used for pulse shaping. And the three normalized SCDs are averaged to obtain smooth curves. At $\alpha = 0$, we can see that FSK signals have the normalized SCD values approaching zero while MSK, PSK and QAM signals have the values greater than zero. Based on these facts, we classify the four modulation types into two types of signals, that is, FSK and non-FSK signals.

For FSK signals, the normalized SCD has discrete spectral lines which are equally spaced over the cycle frequency α [8]. Note that the number of spectral lines is equal to the modulation order. Therefore, we use the number of spectral lines to identify the modulation order for FSK signals.

8.3.2 High-Order Cumulants (HOCs)

HOCs of a signal characterize statistical amplitude features such as the shape of the distribution of the noisy signal [7]. The cumulants has some interesting properties. The cumulant of linearly combined components can be expressed as the sum of the cumulant of each component. Generally, all odd-order cumulants are equal to zero due to the symmetry of symbol constellations. Moreover, HOCs of additive Gaussian noise is also equal to zero.

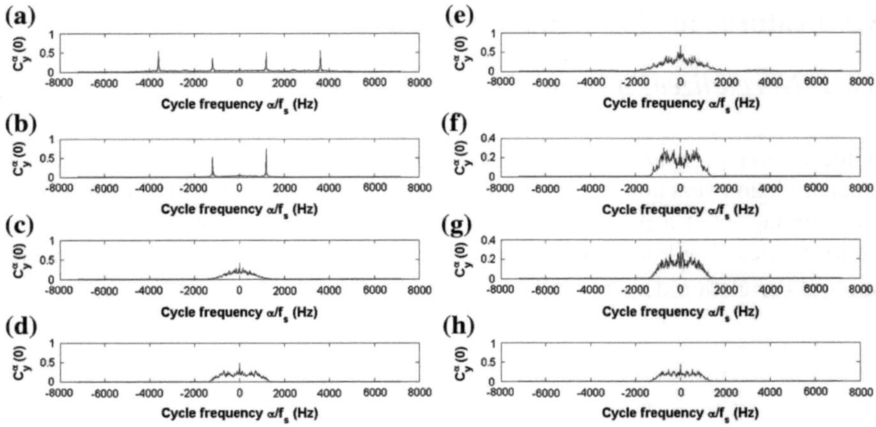

Fig. 8.1 The normalized SCD estimates of **a** 4-FSK, **b** 2-FSK, **c** MSK, **d** BPSK, **e** QPSK, **f** 8-PSK, **g** 16-QAM, and **h** 64-QAM signals

A digitally modulated signal has its inherent shapes of pulse and amplitude distribution. As a result, each modulation signal has distinct even-order cumulants. Therefore, cumulants can be used as features to classify the modulation signal.

For a complex stationary random signal with zero-mean, $y_c(t) = y(t) - E[y(t)]$, zero-lag HOCs used in this paper are expressed as follows [1]

$$\begin{aligned} C_{40} &= cum\{y_c(t), y_c(t), y_c(t), y_c(t)\} = M_{40} - 3M_{20}^3 \\ C_{42} &= cum\{y_c(t), y_c(t), y_c^*(t), y_c^*(t)\} = M_{42} - |M_{20}|^2 - 2M_{21}^2 \\ C_{61} &= cum\{y_c(t), y_c(t), y_c(t), y_c(t), y_c(t), y_c^*(t)\} \\ &= M_{61} - 5M_{21}M_{40} - 10M_{20}M_{41} + 30M_{20}^2M_{21} \\ C_{63} &= cum\{y_c(t), y_c(t), y_c(t), y_c^*(t), y_c^*(t), y_c^*(t)\} \\ &= M_{63} - 9M_{21}M_{42} + 12M_{21}^3 - 3M_{20}M_{43} - 3M_{22}M_{41} + 18M_{20}M_{21}M_{22} \end{aligned} \tag{8.4}$$

where M_{pq} represent the moment of the signal defined as

$$M_{pq} = E[\{y_c(t)\}^{p-q}\{y_c^*(t)\}^q]. \tag{8.5}$$

8.4 The Proposed Modulation Classification Scheme

Figure 8.2 shows the modulation classification flow of the proposed method to classify eight digitally modulated signals of 2FSK, 4FSK, MSK, BPSK, QPSK, 8PSK, 16QAM, and 64QAM. At the first stage of the classifier, we compute the normalized SCD $C_Y^\alpha(0)$ of the signal. The FFT accumulation method of [3] is used to estimate cycle frequencies α. As shown in Fig. 8.1, the normalized SCD at

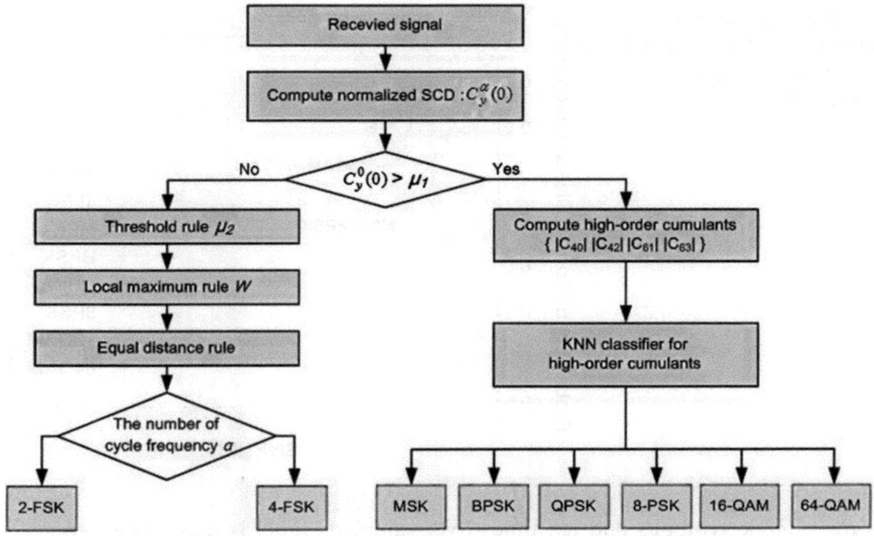

Fig. 8.2 Modulation classification flow of the proposed scheme

frequency $\alpha = 0$ approaches zero for FSK signals. We use a threshold μ_1 to classify FSK and non-FSK signals.

For FSK signals, at the next step, we determine the modulation order with the number of cycle frequencies. The three rules proposed in [8] are used to estimate the cyclic frequencies. First, *a threshold rule* is used to find peaks that are greater than a threshold μ_2. Then, *a local maximum rule* is used to ignore all peaks smaller than a large local peak in the window with a size of W. At the end, *a equal distance rule* is used to find peaks that are equally separated.

For non-FSK signals, we estimate the four HOCs expressed in (8.4) and use KNN classifier. To use KNN classifier, we first create the reference points consisted of absolute values of the four HOCs of non-FSK signals. These reference points form the six groups in the four-dimensional feature space. The four HOCs of the received signal are located as a point in the feature space. We search K nearest reference points, K neighbors, from the point, and predict the modulation group to which maximum number of reference points belongs. Then we decide that the received signal is the corresponding modulation signal.

8.5 Simulation Results

We generate eight modulation signals of 2FSK, 4FSK, MSK, BPSK, QPSK, 8PSK, 16QAM, and 64QAM at baseband. For M-PSK and M-QAM signals, a raised-cosine filter with a roll-off factor of 0.35 is used for pulse shaping. The symbol rate

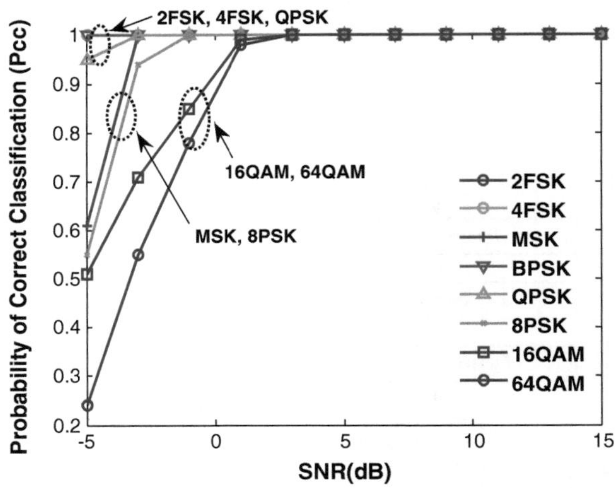

Fig. 8.3 Classification performance of the proposed method for the eight modulation signals

is $R_s = 1.2$ k symbol/s and ten times over-sampled signal is used. The frequency separation is $f_{sep} = 2 \times R_s$ and the sampling rate equals $f_s = 5 \times f_{sep}$ for FSK signals. Each signal is simulated with 100 trials and 14,000 symbols. The signal is normalized to have unit variance. For estimation of the normalized SCD $C_Y^{\alpha}(0)$, we set the resolution of the cycle frequency $\Delta\alpha$ to $f_s/500$ due to the limited observation length 500 samples, and the spectral resolution Δf to $2\Delta\alpha$ in order to satisfy $\Delta f \geq \Delta\alpha$. Also, we empirically set the threshold $\mu_1 = 0.6$ and $\mu_2 = 0.3$ for classification of FSK signals, and the window size $W = 15$.

Figure 8.3 shows the probability of correct classification (P_{cc}) of eight modulation scheme. We see that the classification performance improves as SNR increases. At higher SNR than 3 dB, the correct classification probability approaches 1 with all modulation schemes.

8.6 Conclusion

In this paper, we propose an automatic modulation classification scheme for eight digitally modulated signals of 2FSK, 4FSK, MSK, BPSK, QPSK, 8PSK, 16QAM, and 64QAM, which are widely used in single carrier modulation systems. The normalized SCD and four HOCs are used as features. We use simple thresholds and the KNN algorithm to classify the signals. The simulation results show that the proposed method classifies the signals almost correctly above the SNR of 3 dB.

Acknowledgments This work was supported by the Agency for Defense Development of Korea.

References

1. Aslam MW, Zhu Z, Nandi AK (2012) Automatic modulation classification using combination of genetic programming and KNN. IEEE Trans Wireless Commun 11(8):2742–2750
2. Azzouz EE, Nandi AK (1996) Automatic modulation recognition of communication signals. Kluwer, Boston
3. Da Costa E (1996) Detection and identification of cyclostationary signals. Naval Postgraduate School, Monterey
4. Dobre OA, Abdi A, Bar-Ness Y, Su W (2007) A survey of automatic modulation classification techniques: classical approaches and new trends. IET Commun 1:137–156
5. Gardner WA (1993) Cyclostationarity in communications and signal processing. IEEE Press, New Jersey
6. Like E, Chakravarthy VD, Ratazzi P, Wu Z (2009) Signal classification in fading channels using cyclic spectral analysis. EURASIP J Wirel Commun Networking
7. Nikias CL, Mendel JM (1993) Signal processing with higher-order spectra. IEEE Signal Process Mag 10:10–37
8. Yu Z, Shi YQ, Su W (2003) M-ary frequency shift keying signal classification based-on discrete Fourier transform. In: Military communications conference, MILCOM'03, vol 2. IEEE, pp 1167–1172

Chapter 9
Computing Value at Risk in OpenCL on the Graphics Processing Unit

Nan Zhang, Ka Lok Man and Dejun Xie

Abstract We present our work on computing the value at risk (VaR) of a large hypothetical portfolio in the OpenCL programming model on an AMD FirePro V7900 graphics processing unit (GPU). In the computation of the VaR we follow the delta-gamma Monte Carlo approach. The value change of the portfolio within a short time period is approximated by the sum of a linear delta component and a non-linear gamma component. To approximate the distribution of the value change of the portfolio we generate a large number scenarios. From each scenario a loss or gain of the portfolio is calculated by the delta-gamma approximation. All these potential losses and gains are then sorted, from which an appropriate percentile is chosen as the VaR. We implemented this algorithm in OpenCL. The details are discussed and the experimental results are reported.

Keywords Value at risk · Monte Carlo simulation · Delta-gamma approximation · OpenCL · Graphics processing unit

9.1 Introduction

Value at risk is a simple, easy-to-understand measure for quantifying the potential loss of a portfolio of risky assets within a certain period of time. Because of its multi-faceted advantages the VaR measure has been widely used by financial

N. Zhang · K.L. Man
Department of Computer Science and Software Engineering, Xi'an Jiaotong-Liverpool University, Suzhou, China
e-mail: nan.zhang@xjtlu.edu.cn

K.L. Man
e-mail: ka.man@xjtlu.edu.cn

D. Xie (✉)
Department of Financial Mathematics and Financial Engineering, South University of Science and Technology of China, Shenzhen, China
e-mail: xie.dj@sustc.edu.cn

J.J. (Jong Hyuk) Park et al. (eds.), *Future Information Technology - II*,
Lecture Notes in Electrical Engineering 329,
DOI 10.1007/978-94-017-9558-6_9

practitioners and regulators. The Basel Committee has been using it in setting capital requirements for banks throughout the world [9].

For a given portfolio, an analyst is often interested in making a statement like this: "I am X % confident that the loss of the portfolio will not exceed Y dollars in the next N days." In this statement, the value Y is the VaR of the portfolio, X is the confidence level and N is the time horizon. The VaR says that the loss of the portfolio over the next N days has a probability of only (100 − X) % of being exceeded. Bank regulators require banks to calculate VaR for market risk with N = 10 and X = 99 [8].

To calculate the VaR of a portfolio consisting of non-linear products such as options we follow the delta-gamma Monte Carlo approach. Compared with other methods [4, 5, 11] this approach is a compromise between accuracy and efficiency. According to the Taylor series the loss or gain of the portfolio under discussion can be approximated by a linear delta component and a quadratic gamma component. The distribution of the value change of the portfolio is generated by a Monte Carlo simulation with a large number of trials. An appropriate percentile is then selected from the distribution of the value change to be the VaR of the portfolio. Usually, in calculating the VaR of a portfolio, N is set to 1 day's time. The 10-day VaR is approximately $\sqrt{10}$ times of the 1-day VaR.

To reduce the matrix-matrix multiplication in calculating the quadratic term to matrix-vector multiplication we follow the diagonalisation approach discussed in [6, 7]. The normally distributed variates used in the simulation are generated the from Sobol' sequence [3, 13] computed from Joe and Kuo's direction numbers [10].

For the purpose of accelerating the computation of the VaR for portfolios of thousands of risky assets we attempted on implementing the delta-gamma Monte Carlo approach in the Open Computing Language (OpenCL) programming mode. We decomposed the tasks in the computation and mapped them onto an one-dimensional work space in the OpenCL model. The dependency in the decomposition must be organised in such a way to satisfy the constraint that there is no explicit means to synchronise computations performed on different compute units of the GPU. We tested the OpenCL-VaR implementation on an AMD FirePro V7900 GPU. This model is in AMD's workstation product line. Its maximum power consumption is less than 150 W. It is not designed for superb performance. It has 1,280 stream processors, that is, ALUs. Its core frequency is 725 MHz, and its memory frequency is 1,250 MHz. In the test we used a hypothetical portfolio consisting of 4,000 risky assets. We ran a Monte Carlo simulation with 50,000 trials. It took the GPU 170 s to complete the simulation.

Organisation of the rest of the paper Section 9.2 presents the overall delta-gamma Monte Carlo approach in calculating the VaR of a portfolio and the diagonalisation optimisation. Section 9.3 describes the decomposition of the computation and its mapping onto the work space in the OpenCL model. The section also contains other implementation details about the OpenCL-VaR computation. Section 9.4 presents the test results on the AMD FirePro V7900 GPU. Conclusion and future work are summarised in Sect. 9.5.

9.2 Calculating VaR Using the Delta-Gamma Monte Carlo Approach

We assume the portfolio under discussion consists of N risky assets. The dollar values of these assets at time t are $X_0, X_1, \ldots, X_{N-1}$, respectively. Over a short time period Δt, such as 1 day's time, we assume the returns $\boldsymbol{R} = (\Delta X_0/X_0, \Delta X_1/X_1, \ldots, \Delta X_{N-1}/X_{N-1}) \sim \mathcal{N}^N(\boldsymbol{0}, \Sigma_S)$ of the assets follow a multivariate normal distribution with zero mean vector and the variance-covariance matrix Σ_S. We use $P(t) = \sum_{i=0}^{N-1} X_i$ to denote the aggregated value of the portfolio at time t, $P(t+\Delta t)$ the aggregated value of the portfolio at time $t + \Delta t$, and $\Delta P = P(t+\Delta t) - P(t)$ the value change of the portfolio over the Δt time period. In the delta-gamma VaR approach, this value change ΔP is decomposed into a linear delta component and a non-linear gamma component. The deltas are defined as $\delta_i = \partial P(t)/\partial X_i$, $i = 0, 1, \ldots, N-1$. Each of the deltas measures the sensitivity of the portfolio with respect to the change in value X_i. The gammas are defined as $\Gamma_{ij} = \partial^2 P(t)/\partial X_i \partial X_j$, $i, j = 0, 1, \ldots, N-1$, and each of them measures the sensitivity of the change in δ_i with respect to the change in X_j. The value change $\Delta P = P(t+\Delta t) - P(t)$ is approximated by the Taylor polynomials as

$$\Delta P = \sum_{i=0}^{N-1} \delta_i \Delta X_i + \frac{1}{2} \sum_{i=0}^{N-1} \sum_{j=0}^{N-1} \Gamma_{ij} \Delta X_i \Delta X_j. \tag{9.1}$$

Using the return adjusted deltas and gammas $\tilde{\delta}_i = \delta_i X_i$ and $\tilde{\Gamma}_{ij} = \Gamma_{ij} X_i X_j$, Eq. 9.1 can be written in terms of the returns $R_i = \Delta X_i / X_i$ as

$$\Delta P = \sum_{i=0}^{N-1} \tilde{\delta}_i R_i + \frac{1}{2} \sum_{i=0}^{N-1} \sum_{j=0}^{N-1} \tilde{\Gamma}_{ij} R_i R_j, \tag{9.2}$$

and, in matrix form Eq. 9.2 is

$$\Delta P = \tilde{\boldsymbol{\delta}}^{\mathrm{T}} \boldsymbol{R} + \frac{1}{2} \boldsymbol{R}^{\mathrm{T}} \tilde{\Gamma} \boldsymbol{R}, \tag{9.3}$$

where vector $\tilde{\boldsymbol{\delta}} = (\tilde{\delta}_0, \tilde{\delta}_1, \ldots, \tilde{\delta}_{N-1})$ and matrix $\tilde{\Gamma} = [\tilde{\Gamma}_{ij}]$. The variance-covariance matrix Σ_S and the adjusted gamma matrix $\tilde{\Gamma}$ are symmetric.

Using the Cholesky factorisation the variance-covariance matrix Σ_S can be written as the product of a lower triangular matrix C and C's transpose C^{T}, $\Sigma_S = CC^{\mathrm{T}}$. The correlated returns $\boldsymbol{R} = (R_0, R_1, \ldots, R_{N-1})$ can then be expressed in terms of the independent standard normal variables $\boldsymbol{Z} = (Z_0, Z_1, \ldots, Z_{N-1}) \sim \mathcal{N}^N(\boldsymbol{0}, \boldsymbol{1})$ as $\boldsymbol{R} = C\boldsymbol{Z}$. So, in terms of vector $\boldsymbol{Z}$ Eq. 9.3 becomes

$$\Delta P = \tilde{\boldsymbol{\delta}} C \boldsymbol{Z} + \frac{1}{2} \boldsymbol{Z}^{\mathrm{T}} (C^{\mathrm{T}} \tilde{\Gamma} C) \boldsymbol{Z}. \tag{9.4}$$

To reduce the matrix-matrix multiplications in the gamma term to matrix-vector multiplications, we follow the diagonalisation approach presented in [6, 7]. Using spectral decomposition we can find matrices U and Λ such that $1/2C^{\mathrm{T}}\tilde{\Gamma}C = U\Lambda U^{\mathrm{T}}$, where

$$\Lambda = \begin{vmatrix} \lambda_0 & & & \\ & \lambda_1 & & \\ & & \ddots & \\ & & & \lambda_{N-1} \end{vmatrix}$$

is a diagonal matrix but stored as a vector, and U is an orthogonal matrix ($UU^{\mathrm{T}} = I$) whose columns are eigenvectors of $1/2C^{\mathrm{T}}\tilde{\Gamma}C$. The diagonal elements λ_i of Λ are eigenvalues of $1/2C^{\mathrm{T}}\tilde{\Gamma}C$. We substitute $U\Lambda U^{\mathrm{T}}$ into the gamma term in Eq. 9.4, and it becomes

$$\frac{1}{2}\boldsymbol{Z}^{\mathrm{T}}(C^{\mathrm{T}}\tilde{\Gamma}C)\boldsymbol{Z} = Z^{\mathrm{T}}U\Lambda U^{\mathrm{T}}Z = (U^{\mathrm{T}}Z)^{\mathrm{T}}\Lambda(U^{\mathrm{T}}Z) = \boldsymbol{H}^{\mathrm{T}}\Lambda\boldsymbol{H}, \tag{9.5}$$

where $\boldsymbol{H} = U^{\mathrm{T}}Z$. In order to use vector $\boldsymbol{H}$ in computing the delta term as well, we transform the delta term as

$$\tilde{\boldsymbol{\delta}}C\boldsymbol{Z} = (C^{\mathrm{T}}\tilde{\boldsymbol{\delta}})^{\mathrm{T}}\boldsymbol{Z} = (C^{\mathrm{T}}\tilde{\boldsymbol{\delta}})^{\mathrm{T}}UU^{\mathrm{T}}\boldsymbol{Z} = (C^{\mathrm{T}}\tilde{\boldsymbol{\delta}})^{\mathrm{T}}U\boldsymbol{H} = (U^{\mathrm{T}}C^{\mathrm{T}}\tilde{\boldsymbol{\delta}})^{\mathrm{T}}\boldsymbol{H}. \tag{9.6}$$

If we let vector $\boldsymbol{B} = U^{\mathrm{T}}C^{\mathrm{T}}\tilde{\boldsymbol{\delta}}$, the delta term can be written as $\boldsymbol{B}^{\mathrm{T}}\boldsymbol{H}$. The value change ΔP of the portfolio, therefore, is

$$\Delta P = \boldsymbol{B}^{\mathrm{T}}\boldsymbol{H} + \boldsymbol{H}^{\mathrm{T}}\Lambda\boldsymbol{H} = \sum_{i=0}^{N-1}(B_iH_i + \lambda_iH_i^2). \tag{9.7}$$

In Monte Carlo simulation we generate M N-dimensional vectors $\boldsymbol{Z} = (Z_0, Z_1, \ldots, Z_{N-1})$. Using each $\boldsymbol{Z}$ we calculate a ΔP. We then sort the M ΔP's in ascending order. For a given percentile α, such as 5 %, the 1 day $(1-\alpha)$ VaR is the ΔP at the $\lfloor \alpha M \rfloor$-th position in the sequence.

9.3 Implementing the Simulation in OpenCL

We have implemented the Monte Carlo simulation in OpenCL [12]. In each trial of the simulation a ΔP is calculated using Eq. 9.7. Before the simulation starts, vectors $\boldsymbol{B}$, Λ (conceptually is a matrix) and matrix U are already computed. They remain

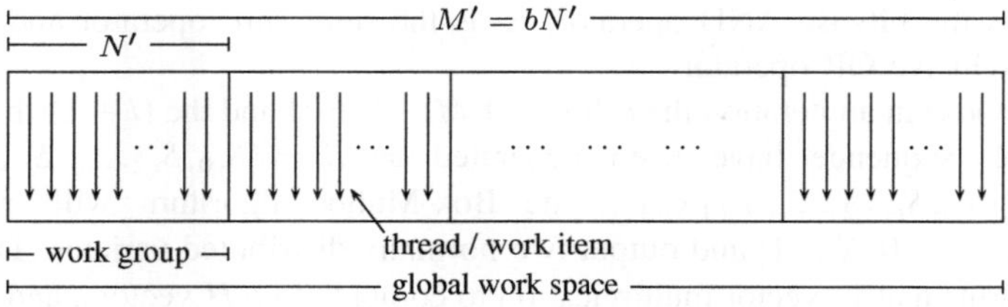

Fig. 9.1 The organisation of the work space and work groups in the OpenCL implementation

constant during the course of the simulation. The programming model [1] of OpenCL requires that we create a work space. The work space consists of a number of work groups. A work group consists of a number of threads, with each thread performing a work item. In our implementation we create a one-dimensional work space with one-dimensional work groups. The structure of the work space is shown in Fig. 9.1.

The work space in total has M' threads. These threads are evenly distributed into N' work groups. To finish M trials for the simulation, each thread performs M/M' trials. For the i-th thread in the work space, that is, the thread whose global id is i, it performs simulation trials within the range $[iM/M',\ iM/M' + M/M' - 1]$. In each trial of the simulation a thread performs the following tasks:

1. Generate a N-dimensional uniformly distributed Sobol' numbers on interval $[0, 1)$.
2. Convert the Sobol' numbers to independent standard normal variates and store them in vector $\boldsymbol{Z}$.
3. Compute vector $\boldsymbol{H} = U^{\mathrm{T}}\boldsymbol{Z}$.
4. Compute the value change ΔP of the portfolio using Eq. 9.7.

Because we use the Box-Muller method [2] to convert two Sobol' numbers to two normal variates, in each iteration a thread actually performs two trials and calculates two ΔP's.

Now we briefly discuss how a N-dimensional Sobol' number is generated by a thread. We assume the thread is working on the i-th trial of the simulation. The input to the Sobol' number generation is a $N(\text{rows}) \times D(\text{columns})$ matrix E storing the pre-computed direction numbers from Joe and Kuo [10]. The number N is the dimension of the Sobol' numbers in the sequence, and D is the number of the direction numbers for each dimension. In our implementation D is set to 32. We use $S_{i,j}$ to denote the j-th, $j \in [0, N-1]$, component of the i-th, $i \in [0, M-1]$ point in the Sobol' sequence, and $E_{j,k}$ to denote the k-th, $k \in [0, D-1]$ number in the j-th row of the direction number matrix E. $S_{i,j}$ is computed as

$$\begin{aligned} S_{i,j} = [((i \wedge 2^0) \gg 0)E_{j,0} \otimes ((i \wedge 2^1) \gg 1)E_{j,1} \\ \otimes \cdots \otimes ((i \wedge 2^{D-1}) \gg (D-1))E_{j,D-1}]/2^{32}, \end{aligned} \tag{9.8}$$

where $\wedge$ is the bitwise AND operator, $\gg$ is the right shift operator and $\otimes$ is the bitwise exclusive OR operator.

We suppose in an iteration the i-th $i \in [0, M-1]$ trial and the $(i+1)$-th points in the Sobol' sequence have been generated as $\boldsymbol{S_i} = (S_{i,0}, S_{i,2}, \ldots, S_{i,N-1})$ and $\boldsymbol{S_{i+1}} = (S_{i+1,0}, S_{i+1,2}, \ldots, S_{i+1,N-1})$. The Box-Muller algorithm will take pairs $(S_{i,j}, S_{i+1,j}), j \in [0, N-1]$ and output two normally distributed variates. The thread then performs matrix-vector multiplication to compute two $\boldsymbol{H}$ vectors, and then two ΔP's. Once all the M ΔP's have been computed the kernel quits. The ΔP's are copied back to main memory, and are sorted. An appropriate percentile is then selected as the 1 day $(1-\alpha)\,\%$ VaR.

9.4 Experimental Result

We have tested the OpenCL-VaR implementation on an AMD FirePro V7900 graphics card (code name "Cayman Pro GL"). It has 1,280 ALUs. Its computing cores run at 725 MHz, and memory runs at 1,250 MHz. As Fig. 9.2 shows, the model has 20 compute units. Each compute unit contains 16 processing elements. Each processing element contains 4 ALUs, which perform integer, single/double-

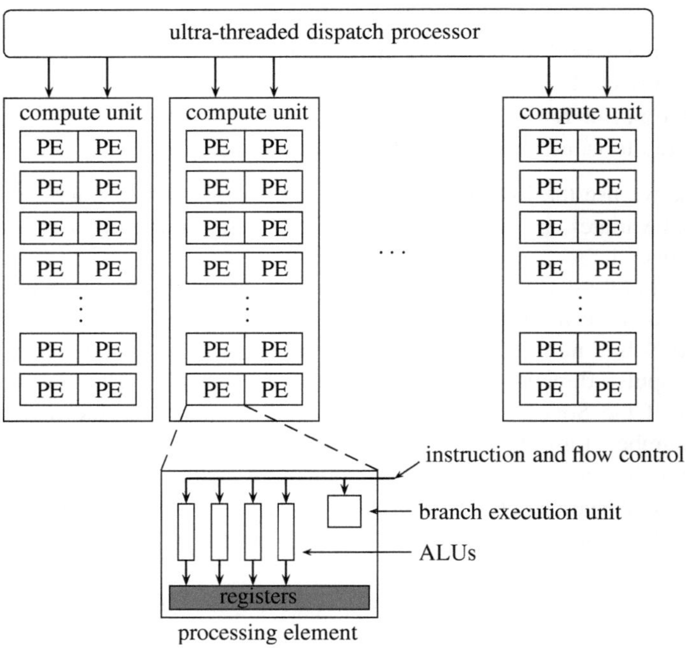

Fig. 9.2 Simplified block diagram of the FirePro V7900 GPU

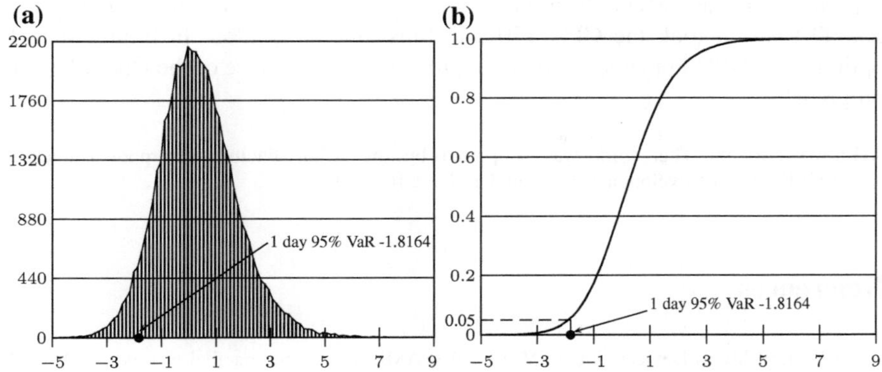

Fig. 9.3 Density and cumulative distribution functions of the portfolio value change. The x-axes of the two figures are both labelled by the loss/gain of the portfolio. The y-axis of **a** is labelled by frequency. The y-axis of **b** is labelled by probability

precision floating point and transcendental operations. Compute units operate independently of each other.

In the test, we setup a hypothetical portfolio of 4,000 risky assets, that is, $N = 4{,}000$. We ran the simulation 50,000 times, thus, $M = 50{,}000$. We set the initial dollar value for each asset to 1. The time period we considered is set to 1 day, so, $\Delta t = 1/365$. It took the GPU 170 s to complete the simulation. The 1 day 95 % value at risk was -1.8164. The largest loss and gain was −4.9 and 9.25, respectively. Between these two extreme values we created 100 equal-sized bins. We counted the number of times the value change fell into each of these bins. The frequency graph is plotted in Fig. 9.3a and the cumulative distribution function of the value change in Fig. 9.3b.

9.5 Conclusions

We have presented our design and OpenCL implementation on the delta-gamma Monte Carlo VaR algorithm. We transferred the Monte Carlo simulation onto the AMD FirePro V7900 GPU for it to perform after the computations before the simulation were completed by the CPU in the same workstation. The organisation of the work space that executes the kernel was a one-dimensional array divided equally into a certain number of work groups. The total number of trials in the simulation were performed by all the threads in the work space with each thread performing an equal number of trials. The number of iterations performed by a thread is half of the number of trials assigned to the thread. This means in each iteration the thread performs two trials because the Box-Muller method takes two uniformly distributed variates and outputs two normally distributed variates. The experiment on the AMD GPU used a portfolio of 4,000 assets and a simulation with 50,000 trials. It took the GPU 170 s to finish the simulation. In future we will optimise the GPU implementation to improve the performance of the OpenCL-VaR program.

Acknowledgments This work was supported by the Xi'an Jiaotong-Liverpool University (XJTLU) Research Development Fund under Grant 10-03-08.

References

1. Advanced Micro Devices, Inc. (AMD) (2013) AMD accelerated parallel processing OpenCL programming guide. http://www.developer.amd.com/appsdk
2. Box GEP, Muller ME (1958) A note on the generation of random normal deviates. Ann Math Stat 29(2):610–611
3. Bratley P, Fox BL (1988) Algorithm 659: implementing Sobol's quasirandom sequence generator. ACM Trans Math Softw 14(1):88–100
4. Britten-Jones M, Schaefer SM (1999) Non-linear value-at-risk. Eur Finance Rev 2:161–187
5. Castellacci G, Siclari MJ (2003) The practice of delta-gamma VaR: implementing the quadratic portfolio model. Eur J Oper Res 150(3):529–545
6. Feuerverger A, Wong ACM (2000) Computation of value-at-risk for nonlinear portfolios. J Risk 3:37–56
7. Glasserman P (2004) Monte Carlo methods in financial engineering, Chap. 9. Springer, New York
8. Hull JC (2007) Risk management and financial institutions, Chap. 8, 1st edn. Pearson Education, New Jersey
9. Hull JC (2012) Options, futures, and other derivatives, Chap. 21, 8th edn. Prentice Hall, New York
10. Joe S, Kuo FY (2008) Constructing Sobol sequences with better two-dimensional projections. SIAM J Sci Comput 30(5):2635–2654
11. Khindanova IN, Rachev ST (2000) Value at risk: recent advances. In: Handbook on analytic-computational methods in applied mathematics. CRC Press LLC, New York
12. Khronos Group: OpenCL (Open Computing Language). http://www.khronos.org/opencl/
13. Sobol' IM (1967) On the distribution of points in a cube and the approximate evaluation of integrals. USSR Comput Math Math Phys 7(4):86–112

Chapter 10
Deserializing JSON Data in Hadoop

Shih-Ying Chen, Hung-Ming Chen, I-Hsueh Chen and Chien-Che Huang

Abstract MapReduce provides an efficient programming framework for processing big data in parallel in Hadoop. On the other hand, as the digitalized data becomes bigger as the advances information of technology, deserializing big JSON data into paths in advance can benefit queries on the data. Therefore, using MapReduce framework to deserialize big JSON data into JSON paths is applicable. In this paper, we propose an efficient JSON data processing mechanism based on MapReduce framework. The mechanism includes a redesign of JSONInputFormat class and the other two Map and Reduce functions.

Keywords JSON · Big data · MapReduce · Hadoop

10.1 Introduction

MapReduce [3, 8, 11] provides an efficient programming framework for processing big data in parallel on Hadoop [2, 9, 10]. The framework is composed of the Map and Reduce phases. During the Map phase, the mappers obtain data from the HDFS and compute them. The results of the mappers are passed to the Reducers to generate the final result in the Reduce phase.

Java Script Object Notation (JSON) [4, 7] is a data-interchange format, which is easy for machines to parse and generate. Being a data-interchange format, JSON is widely used by many applications, such as publishing open data. The basic structures of JSON are name/value pairs and arrays of values. In JSON, an object is an unordered set of name/value pairs, and an array is an ordered set of values. Values are typed as strings, numbers, objects, or arrays.

S.-Y. Chen (✉) · H.-M. Chen · I-HsuehChen · C.-C. Huang
Department of Computer Science and Information Engineering, National Taichung University of Science and Technology, Taichung, Taiwan
e-mail: sychen@nutc.edu.tw

J.J. (Jong Hyuk) Park et al. (eds.), *Future Information Technology - II*,
Lecture Notes in Electrical Engineering 329,
DOI 10.1007/978-94-017-9558-6_10

As to research considering both JSON and MapReduce framework, Jaql [1, 12] processes JSON data in parallel by using Hadoop's MapReduce framework. Jaql is a functional query language that provides users with a simple, declarative syntax for filtering, joining, and grouping JSON data.

In other ways, deserializing JSON data into paths in advance can benefit queries on JSON data. Therefore, using MapReduce framework to deserialize big JSON data into JSON paths is applicable. For example, JSONPath [5, 6] expressions, which refer to JSON structures in the same way as XPath expressions, allow users to query on JSON documents.

In this paper, we propose an efficient data processing mechanism based on MapReduce framework for big JSON data. Different from Jaql, in our proposed mechanism, JSON documents are deserialized into paths for future uses, such as JSON querying, indexing, or mining, etc.

10.2 Analysis and Design

The basic structures of JSON include nested objects and arrays of name/value pairs. In addition, values are typed. In order to deserialize JSON data into paths, information of such structures, including parent-child relationships and types of values, should be recorded. However, a JSON file is fragmented into several splits in MapReduce, the parent-child relationships may be lost after being fragmented. Therefore, a reconstruction of the parent-child relationships among splits during the process of deserializing JSON data into paths is required. This paper proposes a mechanism to deal with JSON data in parallel by the MapReduce framework.

First of all, a JSONInputFormat function is modified from the TextInputFormat class, which is inherited from the InputFormat class in the MapReduce framework. Original TextInputFormat class does not meet the aforementioned requirements. The pseudocode of the JSONInputFormat function is illustrated in Fig. 10.1. The input of the class is a JSON document, and the output is a set of key-valued pairs, where values are name/value pairs in a JSON file and keys are the offsets of the names in the file.

Secondly, we design a MapReduce class to retrieve objects, arrays, and their types of values. The Map function in the class is shown in Fig. 10.2. The function reads from the start of a JSON file to retrieve objects and their positions. It also finds arrays and their positions.

The Reduce function in the class is shown in Fig. 10.3. The function reads the output of the Map function, and reconstructs the parent-child relationships according to the position of objects and arrays by using stacks.

An example of our proposed mechanism is illustrated in Figs. 10.4 and 10.5. In Fig. 10.4, the JSON document includes an array of four objects, one of which has another nested object. The file is fragmented into two splits, split 0 and split 1. Two mappers are invoked by the MapReduce to process the two splits. The mappers retrieve objects and arrays and generate key-value pairs, whose keys are the positions,

```
Input: Path file = split.getPath();
Output:<k1,v1>=<(fk, fv>// Key: offset, Value: name/value pair in JSON
 JSONInputFormat(){
    byte[] separator = ",".getBytes();
    byte[] separator2 = "{".getBytes();
    byte[] separator3 = "}".getBytes();
    byte[] separator4 = "[".getBytes();
    byte[] separator5 = "]".getBytes();
    start = split.getStart();
    end = start + split.getLength();
    if (codec != null) {
       in= new NewLineReader(codec.createInputStream(fileIn), job);
       end = Long.MAX_VALUE;
    } else {
    if (start != 0) {
       skipFirstLine = true;
       this.start -= separator.length;
       fileIn.seek(start);
    }
    if (this.bufferPosn>= this.bufferLength) {
       bufferPosn = 0;
       bufferLength = in.read(buffer);
       if (bufferLength<= 0) {
            break;
       }
    }
    intstartPosn = this.bufferPosn;
    for (; bufferPosn<bufferLength; bufferPosn++) {
       if ( sepPosn> 0  && buffer[bufferPosn] != separator[sepPosn]) {
            sepPosn = 0;
        }
        if (buffer[bufferPosn] == separator[sepPosn]) {
            bufferPosn++;
            int  i = 0;
            for (++sepPosn; sepPosn<separator.length; i++, sepPosn++) {
                 if (bufferPosn + i>= bufferLength) {
                      bufferPosn += i - 1;
                      break;
                 }
                 if (this.buffer[this.bufferPosn + i] != separator[sepPosn]) {
                      sepPosn = 0;
                      break;
                 }
            }
        if (sepPosn == separator.length) {
            bufferPosn += i;
            newline = true;
            sepPosn = 0;
            break;
        }
        bufferPosn--;
    }
    this.bufferPosn=(this.bufferPosn-startPosn)==1?this.bufferPosn:this.bufferPosn-1;
    intreadLength = this.bufferPosn - startPosn;
    bytesConsumed += readLength;
    if (readLength>maxLineLength - txtLength) {
       readLength = maxLineLength - txtLength;
    }
    if (readLength> 0) {
       record.append(this.buffer, startPosn, readLength);
       txtLength += readLength;
       if (newline) {
            str.set(record.getBytes(), 0, record.getLength());
       }
    }
 } while (!newline && (bytesConsumed<maxBytesToConsume));
```

Fig. 10.1 JSONIntputFormat() function

```
Input:<k1,v1>=<(fk, fv> Key: offset Value: name/value pairs in JSON
Output:<k2,v2>=<ak,av>
//replace objects' {}
if (str.indexOf("{") >= 0) {
   xray = true;
   //X's{
   doubleA = "";
   for (int i = CountA; i > 0; i--) {
        doubleA += Countstring+Countnumb+" ";
    }
   //Y's{
    newStr = str.replace("{", "" + Countstring + Countnumb +" "+ doubleA );
        CountB = CountA;
        CountA++;
   }
else if (str.indexOf("}") >= 0) {
   //X's}
    xray = false;
   doubleB = "";
   for (inti = CountB; i > 0; i--) {
        doubleB +=String.valueOf(Countstring)+Countnumb+" ";
    }
   //Y's}
    newStr = str.replace("}", "/" + Countstring + Countnumb +" "+ doubleB );
    CountB--;
    CountA--;

   // next number
   if (CountA == 0) {
        Countnumb++;
   }
 }
// replace arrays' []
if  str.indexOf("[") >= 0  {
   //X[
   doubleC = "";
   for (inti = CountC; i >= 0; i--) {
        doubleC +=Countstring2+(Countnumb2)+" ";
   }
   //Y[
   if(xray){
       newStr = str.replace("[", ""+doubleA+ Countstring + Countnumb+" "+doubleC );
   }
   else
        newStr = str.replace("[", ""+doubleA+ Countstring2 + Countnumb2 +" "+ doubleC
);

   CountD = CountC;
   CountC++;
}
else if (str.indexOf("]") >= 0) {
   //X]
   doubleD = "";
   for (inti = CountD; i >= 0; i--) {
        doubleD +=Countstring2+(Countnumb2)+" ";
   }
//Y]
   if(xray){
       newStr = str.replace("]", "/" +doubleA+ Countstring + Countnumb+" "+ doubleD);
   }
   else
       newStr = str.replace("]","/" +doubleA+ Countstring2 + Countnumb2 +" "+ doubleD
);
   CountD--;
   CountC--;
   //New next number[]
   if (CountC == 0) {
        Countnumb2++;
   }
}
```

Fig. 10.2 Map function

```
Input: <k2,v2>=<(fk1,fk2…fkn), fvi>
Output:<k4,v4>=<aki,avn>
if(key.get() > 0 ){
        stockArr= new String[al.size()];
        stockArr=al.toArray(stockArr);
        stockArr1= new String[a2.size()];
        stockArr1=a2.toArray(stockArr1);
        for(int k=0;k<stockArr.length;k++){
        afterConvert = Integer.parseInt(stockArr1[k]);//key stringtrun to intger
        if(key.get() == afterConvert&& stockArr1[k] != null){
        for(Text val:values){
             stockArr[k].replaceFirst(",","");
             String [] sob =val.toString().split(" ");
             String [] soa = stockArr[k-1].toString().split(", ,")
             int a =soa.length+sob.length;
             String[] soc = new String[a];
             for (int h=0;h<soc.length;h++){
             if (h<soa.length){
                  soc[h] = soa [h];
             }
             else{
                  soc[h] = sob [h-soa.length];
              }
          }
         for (h = 0; h<soc.length;h++) {
             System.out.println(soc[h]);}
             for (i=0;i<soc.length;i++) {
                  if i==soc.length-1 {
                       System.out.println(soc[soc.length-1]);
                       rvalue=stack.push(soc[soc.length-1]);//printed out the last value
                       context.write(newText(stack.prints()),new Text(""));
                       break;
                  }
                  for (j=i+1;j<soc.length;j++){
                       if(soc[j].equals(soc[i])){
                            soc[j] =soc[i] ="";
                            break;
                       }
                       else if j==soc.length-1 &&soc[i] != null {
                            if(soc[i]==""){
                              break;
                            }
                       }
                       rvalue =stack.push(soc[i]);
                       System.out.println("Data"+stack.prints());
                  }
             }
        }
}
```

Fig. 10.3 Reduce function

and values are the objects and arrays. In this example, the array is named "a0" and the objects are named "name" and "big". In Fig. 10.5, the pairs are shuffled to the Reducers, which generate the final results, the deserialized JSON paths. The deserialized paths include "/a0/name/big", "/a0/name", "/a0/name", and "/a0/name", whose values are "A", "b", "c", and "d", respectively.

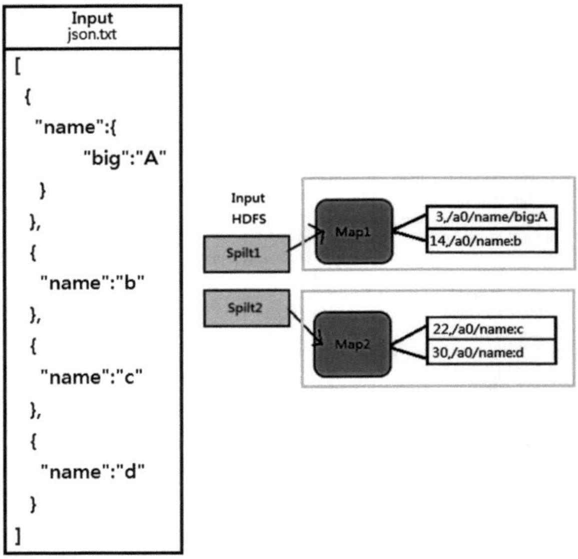

Fig. 10.4 An illustrative example: Map function

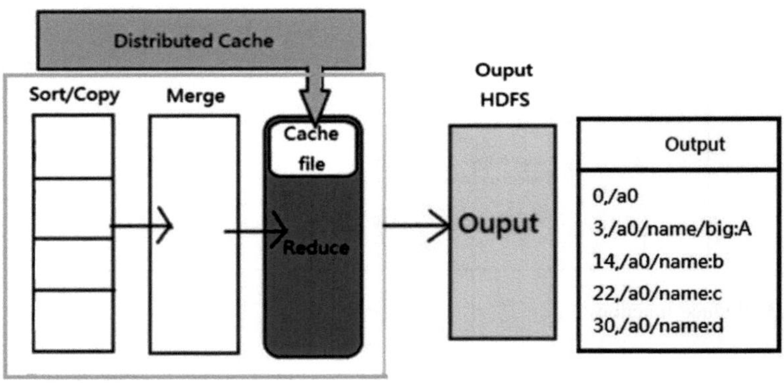

Fig. 10.5 An illustrative example: Reduce function

10.3 Conclusions

Deserializing JSON data into paths in advance can benefit queries on JSON data, especially big JSON data. Therefore, using MapReduce framework to deserialize big JSON data into JSON paths is important. In this paper, we propose an efficient data processing mechanism based on MapReduce framework for big JSON data.

The proposed mechanism reconstructs the parent-child relationships in a JSON file during the deserializing process. The mechanism includes a redesign of JSONInputFormat class and the other two Map and Reduce functions.

References

1. Beyer KS, Ercegovac V, Gemulla R, Balmin A, Eltabakh M, Kanne C-C, Ozcan F, Shekita EJ (2011) Jaql: a scripting language for large scale semistructured data analysis. IBM Research, Almaden, pp 1272–1283
2. Das D HowToHadoop—easy talk about hadoop. http://hadoop4win.nchc.org.tw/cloud/raw-attachment/wiki/HadoopWorkshop/hadoop-assembled.pdf
3. Dean J, Ghemawat S (2008) MapReduce: simplified data processing on large clusters. Commun ACM 51(1):107–113 (50th anniversary issue: 1958–2008 CACM Homepage)
4. Introducing JSON. ECMA-404: the JSON data interchange standard. http://www.json.org/
5. JSONPath. https://code.google.com/p/json-path/
6. JSONPath expressions. http://goessner.net/articles/JsonPath/
7. JSON What it is, how it works, and how to use it; Posted by Jason Lengstorf (in development). http://www.copterlabs.com/blog/json-what-it-is-how-it-works-how-to-use-it/
8. Lee K-H, Lee Y-J, Choi H, Chung YD, Moon B (2011) Parallel data processing with MapReduce: a survey. ACM SIGMOD Rec 40(4):11–20
9. O'Driscoll A, Daugelaite J, Sleatore RD (2013) 'Big data', Hadoop and cloud computing in genomics. J Biomed Inform 46(5):774–781
10. Venner J Pro Hadoop. Apress. http://www.google.com.tw/books?hl=zh-TW&lr=&id=x99mnGARqlYC&oi=fnd&pg=PP1&dq=Inputformat+hadoop&ots=m6shN4ZW9V&sig=d_HsvBiRqKAQO3WHZOB9BW87-9g&redir_esc=y#v=onepage&q=Inputformat%20hadoop&f=false
11. Xiao Z, Xiao Y (2012) Achieving accountable MapReduce in cloud computing. Special issue on extreme scale parallel architectures and systems, cryptography in cloud computing and recent advances in parallel and distributed systems. In: ICPADS 2012 selected papers, vol 30, Jan 2014, pp 1–13
12. Zikopoulos P, Eaton C (2011) Understanding big data: analytics for enterprise class hadoop and streaming data, 1st edn. McGraw-Hill, New York

Chapter 11
Design of a Mobile Brain-Computer Interface System with Personalized Emotional Feedback

Hung-Ming Chen, Shih-Ying Chen, Ting-Jhao Jheng and Shao-Chin Chang

Abstract This study design was based on the emotional evaluation training and instant feedback system on mobile devices linked to brainwave measurement instruments. Through the handheld device software, the evaluation framework for content playback, such as music, videos, photos, etc., was built, and the algorithms for evaluating personalized emotions were developed. The users wore the brainwave measurement instruments to measure the parameters of the brainwaves during the guidance of emotions and to calculate the rates of their emotional influences and trigger keys stored in the database. The evaluation results can be used as the voice or prompt text of the system to provide an immediate personalized emotional feedback response to users.

Keywords Brain-computer interfaces · Mobile devices · Emotion identification · Brainwave feedback

11.1 Introduction

The stress of work, schoolwork, family, economic pressures, and other various factors faced by modern people can easily lead to physical and mental fatigue, panic, insomnia, etc., and can even lead to mental illness in the long term. Listening to music can effectively achieve relaxation and reduce the psychological stress. In the past, traditional research on emotions had no customized services or feedback mechanism for a particular emotion. Relevant studies indicate that music may be effective healing method to influence the music through listening alpha or beta

H.-M. Chen (✉) · S.-Y. Chen · T.-J. Jheng
Department of Computer Science and Information Engineering, National Taichung University of Science and Technology, Taichung, Taiwan
e-mail: hmchen@nutc.edu.tw

S.-C. Chang
Department of Electrical Engineering, Feng Chia University, Taichung, Taiwan

J.J. (Jong Hyuk) Park et al. (eds.), *Future Information Technology - II*, Lecture Notes in Electrical Engineering 329,
DOI 10.1007/978-94-017-9558-6_11

waves [3, 4], however, there is no effective way to know exactly the result of music feedback, so the music feedback will be a research.

Parts of the brain that control the activities have been defined [7] by experts in the field of science of the nervous system. For example, the top of the brain controls the limbs; the rear of the brain is responsible for the control of force, human emotions, mental state, and state of concentration. Different neural activities will produce different brainwave patterns, thus demonstrating a different brain state. Different brainwave patterns make brainwaves of different amplitude and frequency. Beta waves, which are brainwaves between 12 and 30 Hz, mean the brain is in a focused state. Alpha waves, which are brainwaves between 8 and 12 Hz, mean the brain is in a calm, relaxed state.

Regarding human physiological signals and cognitive analysis research, in the 1920s, German neurologist Berger [1] started conducting research on human brainwaves. After decades of research, the results showed that measurement analysis of the brainwaves can show whether or not the user's mental/cognitive status is in a sober state. However, few studies have been able to measure such brainwave signals in real-time. Until recent years, the speed of the computer signal processing technology increased until it could sufficiently process the complex brainwave signals measured in real-time. Information about the mental status thus gradually revealed a research in information science called affective computing.

11.2 Methods

11.2.1 Brainwaves

Brainwave training has already reached the state where people can freely enter different brainwave states. The brainwave segments: delta, theta, alpha, beta, and gamma is shown in Table 11.1.

Because brainwave feedback training is a complex training pattern that aims to let people experience a specific state of consciousness and how to enter these states. For example, a sound or visual message can be used to create a personal feedback mode, and this requires implementing personal tasks many times to achieve the most effective feedback mode.

Table 11.1 Each brainwave frequency segment

Brainwave type	Frequency range (Hz)
Delta δ	1–3
Theta θ	4–7
Alpha α	8–9 (low), 10–12 (high)
Beta β	13–17 (low), 18–30 (high)
Gamma γ	31–40 (low), 41–50 (mid)

11.2.2 Brain-Computer Interface Control

11.2.2.1 The Overview of Brainwave Sensing Device

Usually, brainwaves are measured by trained personnel who place many measuring electrodes, as well as gel, on the head. This traditional measurement method is cumbersome and often leads to incorrect results or interruptions that delay each measurement for a long time. NeuroSky in 2009 released the first Brain-Computer Interfaces (BCI) control. Neurosky MindSet is different from the past BCI devices. It is a single EEG channel comparing with the traditional multi-channels EEG technology and makes wearing easier and more comfortable with no need of people to help you wear it. The NeuroSky product uses its patented ThinkGear chip that can filter out the noise from the EEG to get eSense [5].

Traditional brainwave measuring instruments are only used to display the results and analysis on the computer and had low portability and device durability. We combined the convenience, durability, and portability of mobile phones with the brain-computer interface (BCI) technology of Neurosky's brainwave headset (Mindset) to carry out real-time acquisition and analysis of EEG data and send the current attention, meditation, and other values [6] to a smartphone with Bluetooth wireless technology, so as to easily interpret the current state of mind.

Our way of focus can cause a measurable shock to brainwaves, which is recorded [2] by EEG. Scientists in the application of EEG confirmed that different forms of focus will form different brainwave frequencies. Brainwave activities are usually recorded in the range of 1–50 Hz. The so-called brainwave refers to the electrical activity of the brain. Current activities at all times are recorded and converged into an electric current pulse called the brainwave, which can be measured through the forehead (called FP1 area in neuroscience) using brainwave headset sensors.

Therefore, this system measured the users' brainwave state while they were listening to music or watching videos, photos, and so on, and scored their current state. Through the scores, users can understand what kind of music, movie, or photo has the greatest influence on their emotion. In addition, through the manner of real-time voice or text, the relative voice or text is given to the user's emotion, enabling the user to adjust their emotional state through voice. The schematic diagram of the overall system is shown in Fig. 11.1.

11.2.3 Brainwave Feedback of Emotional Control System

11.2.3.1 System Overview

With Android smart phone, we achieved emotional control and adjustment and emotional effect rating. The brainwave sensing headset should be connected to the

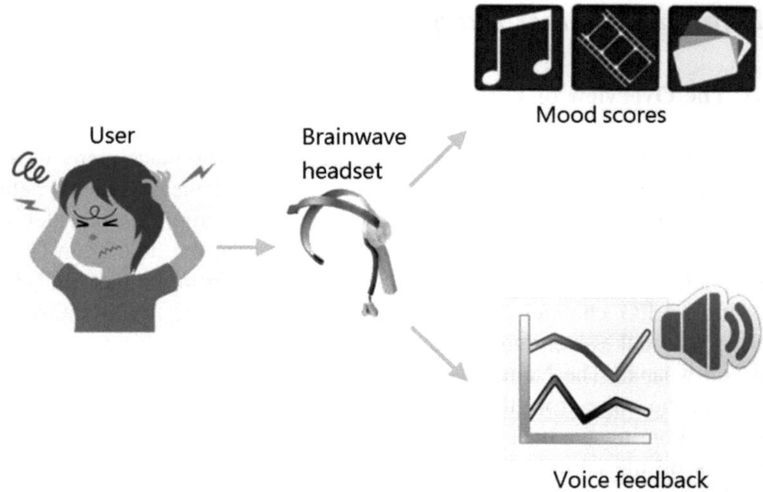

Fig. 11.1 The applied schematic diagram of wearing EEG measurement headsets

Android smart phone via Bluetooth to apply this system, including the training mode and real-time mode. Refer to Fig. 11.1 for the applied schematic diagram of wearing brainwave sensing headsets, and Fig. 11.2 for the whole screen of the system.

The sequential diagram is used to illustrate the sequential relationship of calls and returns between programs or modules. For example, Figs. 11.3 and 11.4 show the training mode sequential diagram of the brainwave feedback for emotional control system: The content called the brainwave data access unit is followed by the user conducting the entertainment unit and then the mood scores and record storage.

11.2.3.2 Training Mode

For the screen composed of four units, refer to the training mode interface diagram in Fig. 11.5a. The content of each entertainment unit includes a music subsystem, film subsystem, photo subsystem, and database subsystem. The entertainment unit content contrasts the general users' relaxation approaches and at the same time conducted brainwave data acquisition and analysis. After the analysis, the emotional impact score is displayed and stored in the database subsystem. A part of the algorithm is described in Sect. 11.2.3.4. The database subsystem is described in Sect. 11.2.3.5.

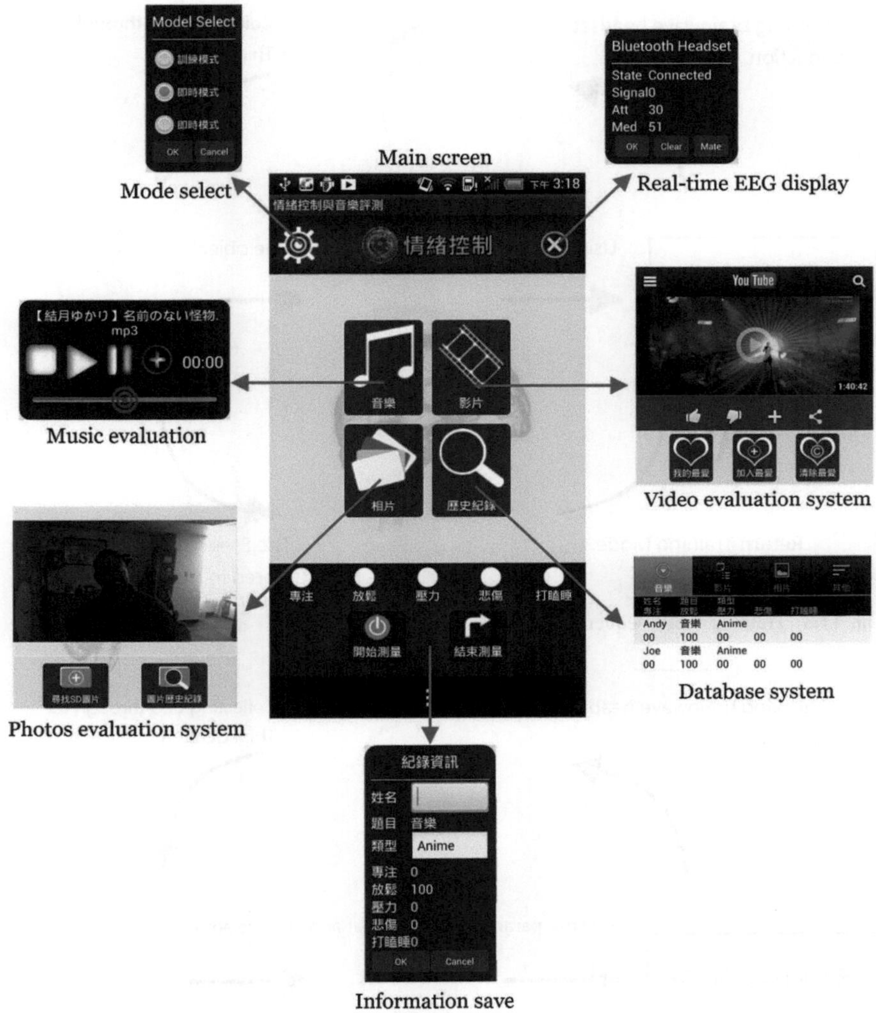

Fig. 11.2 System operation interface diagram

11.2.3.3 Real-Time Mode

The content uses a line chart to show the user's current attention (in red) and Meditation (in blue). Refer to Fig. 11.5b for the real-time mode interface diagram. In addition to the brainwave line chart, this mode can also produce a voice every 10 s to remind users to adjust their emotional state. A part of the algorithm is described in Sect. 11.2.3.4.

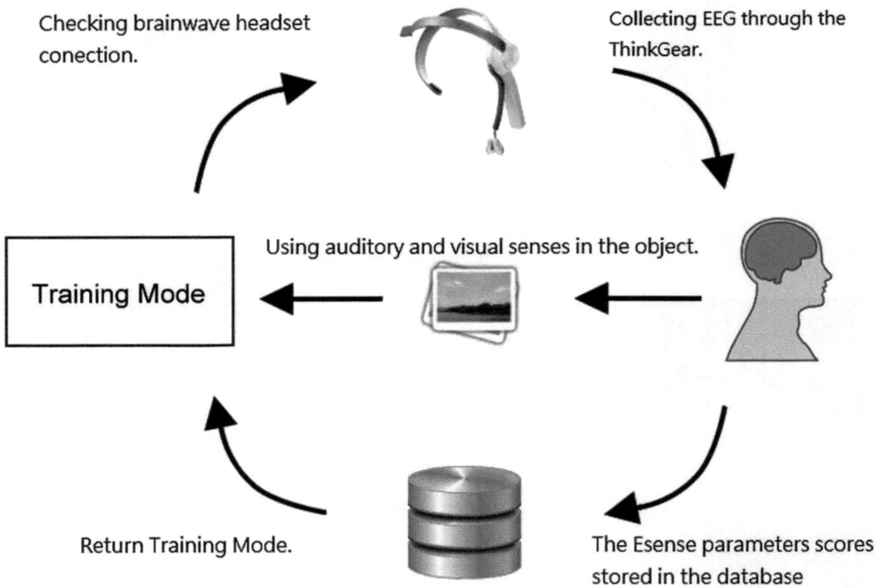

Fig. 11.3 Training mode sequential diagram

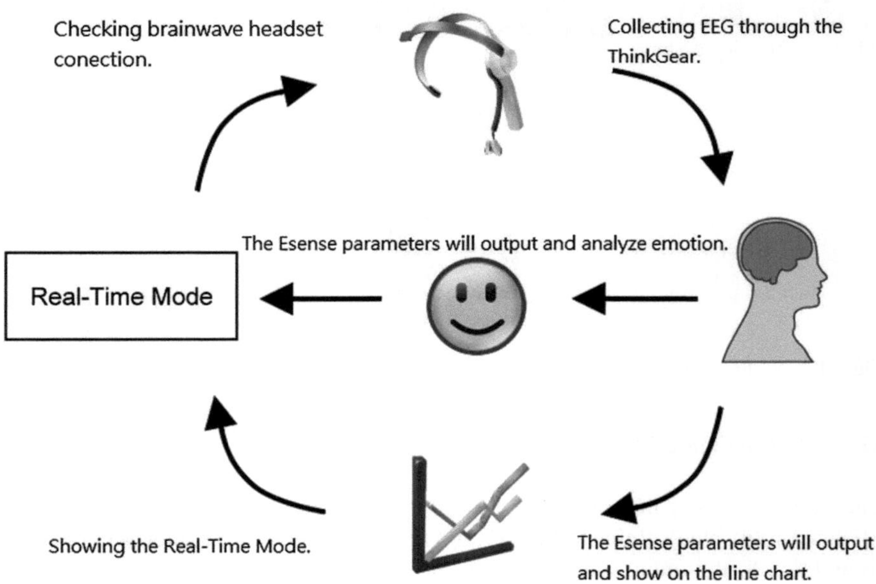

Fig. 11.4 Real-time mode sequential diagram

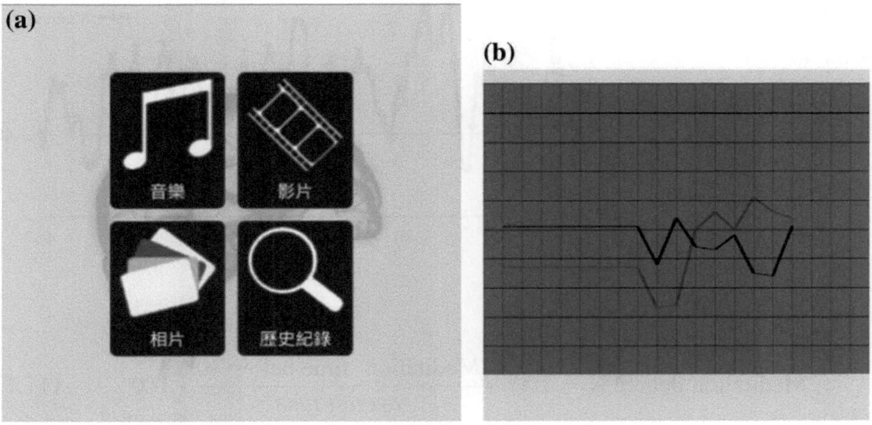

Fig. 11.5 **a** Training mode interface diagram. **b** Real-time mode interface diagram

11.2.3.4 Algorithm Design

The phenomenon with electrical shock characteristics and electrical wave generated in general human brain cell activity is called a brainwave. Recent medical studies have proven that various activities in the human brain, including thoughts, emotions, desires, etc., are revealed by the current and chemical reactions, and waveforms of different vibration frequencies can be measured by brainwave instruments. In studies by Crowley et al. [8], the intensities of Meditation degree and attention degree were calculated. Refer to Fig. 11.6 for the emotional interval determination conditions. Therefore, this study refers to the design of emotional scoring algorithm, converted into a score from 0 to 100, as shown in Eqs. (11.1)–(11.2).

11.2.3.5 Database Subsystem

As a result of storing user scorings, which will help further analysis, the database field design is as shown in Table 11.2.

Table 11.2 Database field design

Field name	Example
_id(PK)	1
name(not null)	Ting-Jhao Jheng
object(not null)	Music subsystem
object_name(not null)	Andy Lau-desensitizing water
attention_score	98
Meditation_score	64

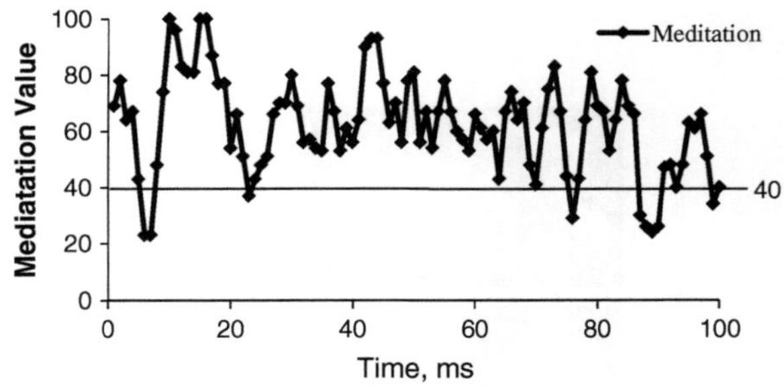

Fig. 11.6 Cumulative line chart with meditation degree below 40

$$\text{Meditation degree X} = 100 - \frac{\text{Meditation time below 40}}{\textit{overall time}} * 100 \tag{11.1}$$

$$\text{Attention degree X} = 100 - \frac{\text{Attention time below 40}}{\textit{overall time}} * 100 \tag{11.2}$$

11.3 Conclusions

The ultimate goal of this study is to recommend to each user a personalization system, which can help adjust emotions according to the different needs and also adjust the current mood through the real-time mode. The system development can be used as a framework, to introduce the essential oil case to adjust the emotional state in the future. The brainwave sensing headphone was used to combine application for the development of the emotional control system, in order to feedback emotion to the user in a way different from the past. This research will continue to conduct the experiments and analysis of focus and relaxation and to achieve personal emotional adjustment.

References

1. Berger H (1929) Uber das Elektroenkephalogram des Menschen. In: AfPN, vol 87. Springer, Berlin, pp 527–570
2. Crowley K, Sliney A, Pitt I, Murphy D (2010) Evaluating a brain-computer interface to categorise human emotional response. In: 2010 IEEE 10th international conference on advanced learning technologies (ICALT). IEEE Press, Sousse, pp 276–278
3. DeSteno D, Petty RE, Rucker DD, Wegener DT, Braverman J (2004) Discrete emotions and persuasion: the role of emotion-induced expectancies. J Pers Soc Psychol 86:43–56
4. Hassan H, Murat ZH, Ross V, Buniyamin N (2012) A preliminary study on the effects of music on human brainwaves. In: 2012 international conference on control, automation and information sciences (ICCAIS). IEEE Press, Ho Chi Minh City, pp 176–180

5. NeuroSky, Inc. (2009) NeuroSky's eSense™ meters and detection of mental state. Technical report, NeuroSky, Inc.
6. Parimal Kumar KR, Prasanna Kumar KS, Thakur MK (2013) Brain wave as an input device-new approach to control digital devices. In: International conference on management and information systems, vol 22
7. Popper KR, Eccles JC (1977) The self and its brain. Springer, Berlin
8. Crowley K, Sliney A, Pitt I, Murphy D (2010) Evaluating a brain-computer interface to categorise human emotional response. In: 2010 IEEE 10th International Conference on Advanced Learning Technologies (ICALT). IEEE Press, Sousse, pp 276–278

Chapter 12
Efficient XML Data Processing Based on MapReduce Framework

Shih-Ying Chen, Hung-Ming Chen and Wei-Chen Zeng

Abstract Due to the advances of information technology, new devices generate amount of data. Especially, XML is a standard format for data exchange. Therefore, processing big XML data is an important topic. We propose an efficient XML data processing mechanism, which includes a design of XMLInputFormat class, MapReduce modules, and an HBase schema. The mechanism scans an XML document to reconstruct parent-child relationships in the document. It generates deserialized paths, which are stored in HBase.

Keywords XML · Big data · MapReduce · Hadoop

12.1 Introduction

In recent years, due to the advances of information technology, new devices generate amount of data, such as mobile phones and sensors. As a result, processing such big data becomes an important topic. Cloud computing technology is useful to achieve the purpose. For example, [5] provides efficient search result by a cloud computing framework consisting of thousands of computers.

In order to process big data, MapReduce framework [2, 4, 9, 11] is adopted. MapReduce, which works on the Hadoop Distributed File System (HDFS) [10], is one of the most important data processing frameworks. It divides the computing process into the Map and Reduce phases. The input data of the mappers is a set of

S.-Y. Chen (✉) · H.-M. Chen · W.-C. Zeng
Department of Computer Science and Information Engineering, National Taichung University of Science and Technology, Taichung, Taiwan
e-mail: sychen@nutc.edu.tw

H.-M. Chen
e-mail: hmchen@nutc.edu.tw

W.-C. Zeng
e-mail: s18023105@nutc.edu.tw

J.J. (Jong Hyuk) Park et al. (eds.), *Future Information Technology - II*,
Lecture Notes in Electrical Engineering 329,
DOI 10.1007/978-94-017-9558-6_12

key-valued pairs, denoted by $<key_1, value_1>$. The Mappers process the pairs to generates another set of pairs, denoted by $<key_2, value_2>$. $<key_2, value_2>$ pairs of a Mapper are hashed into region files by a hash function of key_2. Therefore, the pairs of the same value of key_2 are hashed into the same region file. A Reducer then reads the hashed pairs from its own region files, and merges the pairs to generate a list of pairs, denoted by $<key_2, list(value_2)>$. The pairs are further computed to generate the final result in each Reducer.

In the MapReduce framework, distributed cache [7] can be used to distribute read-only data files. It is a facility to cache files needed by MapReduce applications. It increases the efficiency of applications if the cached data is not too large.

For storing data in the cloud, HBase [6] is a widely-used distributed storage system. HBase provides random, real time read/write access to big data. In contrast with traditional relational databases, HBase is a column-based database, which can store very large tables. It provides APIs for programmers to read/write HBase tables.

Extensible Markup Language (XML) is a standard format for data exchange. It is used for handling all types of complex documents, supporting information inquiries, electronic data interchange and other applications. Many studies focus on efficient XML indexing scheme [1, 3, 8], XML parsing, XML data mining, etc. However, these XML data processing techniques do not consider how to cope with big XML data.

In this paper, we propose an efficient data processing mechanism based on MapReduce framework for big XML data. We take advantages of MapReduce framework and the distributed cache technique. In the proposed mechanism, XML documents are deserialized into paths, which are stored in HBase [6] for further uses, such as XML querying, indexing, or mining, etc.

12.2 Proposed Method

An XML document consists of data content and structure. The structure consists of nested tags and forms parent-child relationships. For example, there are four tags in the XML document in Fig. 12.1. They form an XML tree. However, during the

Fig. 12.1 An XML document

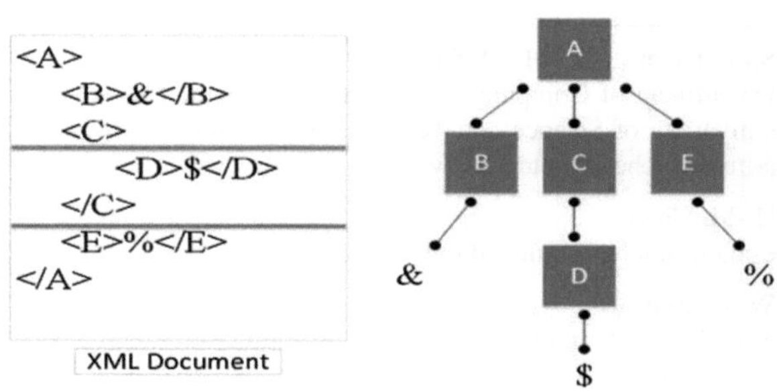

processing of MapReduce, an XML file is fragmented into splits. As a result, the parent-child relationships among tags are damaged. In this paper, we propose a method to reconstruct parent-child relationships among the splits in order to deserialize XML documents into paths, which will be stored in HBase. The deserialized paths are useful for XML querying, indexing, mining, etc. The proposed mechanism includes four parts: a design of XMLInputFormat class, two MapReduce modules, and an HBase schema design.

We design an XMLInputFormat class, which inherits the InputFormat class in MapReduce, for the purpose of processing big XML data. Figure 12.2 illustrates the pseudo code of the class. The XMLInputFormat function in Fig. 12.2 copes with two main goals. The first one is records the tags' starting offsets in an XML document as well as their contents. The second one is that the content of each tag must be complete in the function.

The input of the XMLInputFormat function is an XML file. The output is a set of key-valued pairs, whose keys are starting offset of tags and values are tag names as well as their data content. Since the output records the positions where tags start in an XML document, the Mappers can take advantage of the information.

In addition, since the aforementioned splits may start at any file position, tags as well as their contents may thus be separated between splits. For example, Fig. 12.1 illustrates the three splits of the document. Therefore, to achieve the second goal, the XMLInputFormat function continues scanning the texts in the next split if necessary in order to retrieve a complete content data of the processing tag. For example, in Fig. 12.1, if the file is fragmented into two splits at the position before the string "$", in order to retrieve a complete tag as well as its content, the XMLInputFormat function continues scanning the texts to the end tag "</D>" even if the tag "</D>" is in the second split.

We reconstruct the damaged parent-child relationships of the aforementioned splits by two MapReduce modules. The first one scans each split to get elements that are across splits. The second one reconstructs the parent-child relationships.

The pseudo code of the first module is shown in Fig. 12.3. The module records the elements that are across splits and stores the information into distributed cache. The input of the module is the output of the XMLInputFormat function, and the output of the module is the positions of the start tags that are across splits.

The process of Fig. 12.3 is illustrated in Fig. 12.4. In Fig. 12.4, the XML document is fragmented into splits 0–2. Each split is processed by an XMLInputFormat function to generate the output. The output of the module is a set of key-valued pairs, whose keys are tag names and values are the positions of the tags. They are <0, <A>>, <11, <C>>, <22, </C>>, <34, </A>>, since the elements of <A> and <C> are across splits. The numbers of the pairs are the positions of the tags. The output is stored in a distributed cache.

The pseudocode of the second module is shown in Figs. 12.5 and 12.6. Figure 12.5 shows the Map function. The input of this function is an XML document. The Mappers read the splits of the XML document to generate key-valued pairs, whose keys are the positions, and values are tags as well as their content texts. The pairs are shuffled into Reducers, whose pseudocode is shown in Fig. 12.6.

```
Input: Path file = split.getPath();
Output:<k1,v1>=<fkn, fvi>//fkn、 fvito XMLInputFormat cutting out Key: String
offset, Value: <tag> and <content_text>
XMLInputFormatclass{
XMLInputFormat function{
    Byte[] Start separator = "</".getBytes();
    Byte[] End separator = ">".getBytes();
    this.maxLineLength = job.getInt("mapred.linerecordreader.maxlength",
    Integer.MAX_VALUE);
    start = Start position;
    end = End position control slice;
    if start ! = 0 {
    this.start - = separator.length;
    fileIn.seek(start);
    }
    fvi = start;
while pos< end {
    newSize = in.readLine(value, maxLineLength,
    Math.max((int) Math.min(Integer.MAX_VALUE, end - pos), maxLine-
Length));// Read the contents of a line
    pos += newSize;
    if newSize smaller than the file size{
        break;
    }
    for bufferPosn< buffer.Length-1 {
        if sepPosn> 0 && buffer[bufferPosn] != Start separator [sepPosn] {
            sepPosn = 0;
        }
        if buffer[bufferPosn] == Start separator [sepPosn] {
            sepPosn = 0;
            bufferPosn++;
            Set the value of i to 0;
          for (++sepPosn; sepPosn< Start separator.length; i++, sepPosn++) {
              if bufferPosn + i>= bufferLength {
              Over an array of values ;
              bufferPosn += i - 1;
              break;
              }
          if this.buffer[this.bufferPosn + i] != Start separator [sepPosn] {
                sepPosn = 0;
                break;
              }
          }
        if sepPosn == Start separator.length {
            while this.buffer[this.bufferPosn] != End separator[0] {
                bufferPosn++;
              if(bufferPosn>this.buffer.length-1){
                bufferPosn-=2;
                sepPosn = 0;
                break;
                }
            }
            bufferPosn++;
            newline = true;
            sepPosn = 0;
            readLength = this.bufferPosn - startPosn;
            bytesConsumed += readLength;
          if (newline) {
                fvi =str.set(record.getBytes(), 0, record.getLength());
                Output:<k1,v1>=< fkn, fvi >;
          } } } } }}
```

Fig. 12.2 XMLInputFormat class

```
Input:<k1,v1>=<fk_n, fv_i> Key: String Offset, Value: <tag> and <content_text>
Output:<k2,v2>=<ak_i,av_i>//aki, avi string offset the parent class of XML tags
Map class{
    Map function{
        Stack[ ]=new Stack [ ];
        k2 is Set the value of k1;
        Tagname is an array storing v1.tags;
        For i = 0 to Tagname.length-1{
            If (i == Tagname.length-1){
                If Tagname != null {
                    Stack.push(Tagname[i]);
                }
            }
            For j =i+1 to Tagname.length-1{
                If values are equal {
                    Tagname Intermediate value less;
                    break;
                }else if j == Tagname.length-1{
                    Stack.push(Tagname [i]);
                }
            output<k2,v2>=<[ak1,av1][ak2, av2]...[aki,avi]>;
} } } }
```

Fig. 12.3 Pseudocode of the first module

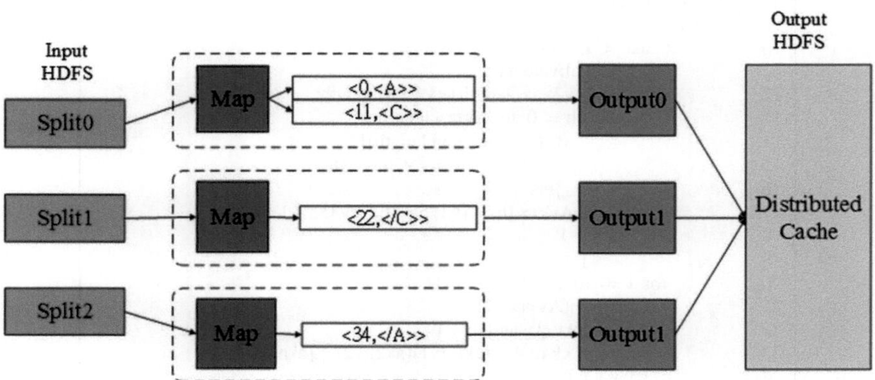

Fig. 12.4 Illustration of the first module

In Fig. 12.6, the input of this function is the output of the Map function, and the output of the function is the deserialized XML paths. In the phase, by using the key-valued pairs from the Mappers and the distributed cache, the final result, i.e. deserialized XML paths, is generated.

The process of the second module is illustrated in Fig. 12.7. In Fig. 12.7, the Mappers read the three splits of the XML document to generate key-valued pairs,

```
Input: <k1,v1>=<fk_n, fv_i> Key: String offset, Value: <tag> and <content_text>
Output:<k3,v3>=<ak_i,rvi>
Map class{
    Map function{
    Stack[] = new Stack[ ];
      k3 is Set the value of k1;
      Tagname is an array storing v1.tags;
      Tagname is value Cutting;
     For i=0 to Tagname-1{
          Stack.push(Tagname);
       If Encounters the end tag {
           output <k3,v3>=<[ak1,av1][ak2, av2]...[aki,avi]>;
       } } } }
```

Fig. 12.5 Pseudocode of the Map function in the second module

```
Input:<k3,v3>=<ak_i,av_i>
Output:<k4,v4>=<ak_i,rvn>,where aki string that represents the offset and Value:
String offset pursuant to distinguish <tag> sort and merge
Reduce class{
  Setup(Context context) function{
    Read <k2,v2>=<ak_i,av_i> into distributedcacahe;
    StringTokenizertokenizer = new StringTokenizer(line," ");
    ak_i =tokenizer.nextToken().trim();
    av_i = tokenizer.nextToken();
  }
  Reduce function{
    k4 is Set the value of k3;
    if k2.aki < k3.aki{
      for Text val:values{
        Av<t> = <v2>avi.length+<v3>avi.length;
          for h = 0 to Av<t>.length-1{
            if (h < <v2>avi.length){
              Av<t> [h] = <v2>avi. [h];
            }else{
            Av<t> [h] = (v3) avi [h - <v3>avi.length];
            }
          }
      for i =0 to Av<t>.length-1{
        stack.push(Av<t> [i]);
        if i == Av<t>.length-1 {
          output <k4,v4><[ak_1, rv1][ak2,rv2]...[akn, rvn]>;
        } } } } } }
```

Fig. 12.6 Pseudocode of the Reduce function in the second module

whose keys are the positions, and values are tags as well as their content texts. The pairs are shuffled into Reducers. The Reducers use the pairs and read the distributed cache to generate the final result, the deserialized XML paths.

The deserialized XML Paths are stored into HBase for further uses. Our proposed HBase schema is shown in Fig. 12.8. Based on the schema, the HBase table of XML document in Fig. 12.1 is illustrated in Fig. 12.9. Users can utilize the deserialized XML paths to index, query, or mine XML documents.

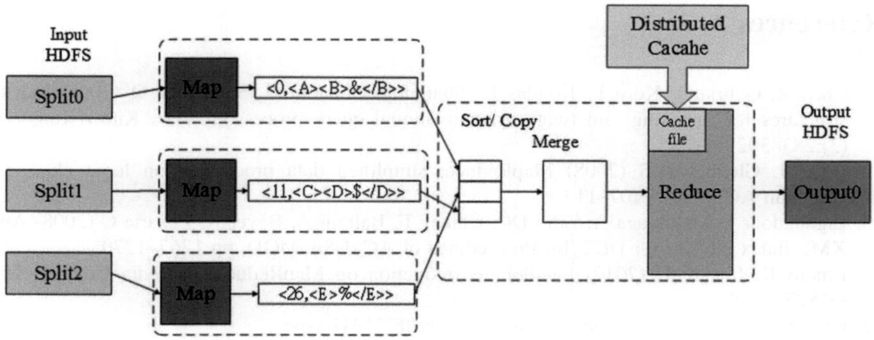

Fig. 12.7 Illustration of the second module

row_key	Column family: *cf1*
XML_path	Column qualifier: *context_text*

Fig. 12.8 HBase schema for storing deserialized XML paths

row_key	Column family: *cf1*
'/A/B'	*context_text= '&'*
'/A/C/D'	*context_text= '$'*
'/A/E'	*context_text = '%'*

Fig. 12.9 HBase table storing the deserialized XML paths

12.3 Conclusions

In this paper, we propose an efficient data processing mechanism based on MapReduce framework for big XML data. The proposed mechanism includes four parts: a design of XMLInputFormat class, two MapReduce modules, and an HBase schema design. The mechanism reconstructs parent-child relationships among the MapReduce file splits and deserializes an XML document into paths, which are stored in HBase for further uses. The result of our research can utilize XML querying, indexing, mining, etc.

Acknowledgments This work is supported in part by NSC in Taiwan, R.O.C. under Grant No. NSC-100-2221-E-025-014. The authors are also grateful to the reviewers for their helpful comments.

References

1. Chen Z, Gehrke J, Korn F, Koudas N, Shanmugasundaram J, Srivastava D (2007) Index structures for matching xml twigs using relational query processors. Data Knowl Eng 60 (2):283–302
2. Dean J, Ghemawat S (2008) MapReduce: simplified data processing on large clusters. Commun ACM 51(1):107–113
3. Elghandour I, Aboulnaga A, Zilio DC, Chiang F, Balmin A, Beyer K, Zuzarte C (2008) An XML index advisor for DB2. In: Proceedings of ACM SIGMOD, pp 1267–1270
4. Emoto K, Imachi H (2012) Parallel tree reduction on MapReduce. Procedia Comput Sci 9:1827–1836
5. http://msdn.microsoft.com/zh-tw/windowsazure/ff721941.aspx
6. http://oss-tw.blogspot.tw/2010/04/hbase-vs-cassandra.html
7. https://hadoop.apache.org/docs/r1.2.1/api/org/apache/hadoop/filecache/DistributedCache.html
8. Lin X, Wang N, Xu D, Zeng X (2010) A novel XML keyword query approach using entity subtree. J Syst Softw 83:990–1003
9. Roddick C, Braganholo V, Mattoso M (2011) Virtual partitioning ad-hoc queries over distributed XML databases. J Inf Data Manage 2:495–510
10. White T (2010) Hadoop: the definitive guide. O'Reilly Media, Sebastopol
11. Zinn D, Bowers S, Kohler S, Ludascher B (2010) Parallelizing XML data-streaming workflows via MapReduce. J Comput Syst Sci 76:447–463

Chapter 13
Examining Organizational Value Co-creation Behavior from the Perspective of Service Science

Huan-Ming Chuang and Chyuan-Yuh Lin

Abstract S-D logic is a more robust framework for service science than the traditional goods-dominant (G-D) logic that emphasizes the importance of operant resources and value co-creation. This study examines the dynamic nature of organizational value co-creation behavior by DEMATEL-based ANP. Major research findings can be expected to provide helpful guidance for the effective promotion of organizational value co-creation behavior to leverage the operant resources to their maximum potentiality.

Keywords Service dominant logic · Value co-creation · DEMATEL-based ANP

13.1 Introduction

Academics and practitioners both notice the important relationships among service, performance, and competitive advantage. The link between superior performance and competitive advantage has been well validated. Yet according to Lusch et al. [17], there is little evidence of significantly increasing service; therefore, they posit that a relatively new perspective, named service dominant (S-D) logic is needed to view exchange, markets, and competing through service.

S-D logic is a more robust framework for service science than the traditional goods-dominant (G-D) logic [18], with the following important tenets: (1) service is conceptualized as a process, rather than a unit of output, (2) service focuses on dynamic (or operant) resources, such as knowledge and skills, rather than static (or operand) resources, such as natural resources, and (3) service looks upon value as a

H.-M. Chuang · C.-Y. Lin (✉)
Department of Information Management, National Yunlin University of Science and Technology, 123, University Road, Section 3, Douliou 64002, Yunlin, Taiwan, ROC
e-mail: g9923807@yuntech.edu.tw

H.-M. Chuang
e-mail: Chuanghm@yuntech.edu.tw

J.J. (Jong Hyuk) Park et al. (eds.), *Future Information Technology - II*,
Lecture Notes in Electrical Engineering 329,
DOI 10.1007/978-94-017-9558-6_13

collaborative process between stakeholders (i.e., providers and customers) rather than producers create then deliver to customers.

Under this background, organizational performance and competitive advantage hinge on the effectiveness of its value co-creation network formed by participatory employees. Since knowledge and skills are important operant resources under S-D logic, knowledge sharing behavior is one of essential value co-creation activities that deserve deep examination. Knowledge sharing between employees and within and across teams allows organizations to exploit and capitalize on knowledge related resources [8, 11] to create and sustain competitive advantages in a highly competitive and dynamic economy [11, 13].

Because knowledge sharing behavior is occurring in social process, besides technology investments, other social and environmental factors are required for its final success. Consequently, social-technical approach is appropriate to help us understand the way in which technology is adapted and used in an organization [7].

As a result, this study follows social cognitive theory to first identify important elements related to environmental, personal, and behavioral dimensions from literature review. Then appropriate tools from multi criteria decision making (MCDM), such as DEMATEL and ANP are applied to explore the triple reciprocity of value co-creation behavior.

13.2 Theoretical Background

This study bases on social cognitive theory to explore the triple reciprocity of organizational value co-creation behavior. Related theories are discussed below.

13.2.1 Social Cognitive Theory

Social cognitive theory (SCT) is a widely accepted, as well as empirically validated model of individual behavior [10]. It is based on the premise that cognitive and other personal factors, environment influences, and behavior are reciprocally determined [26].

In general, personal factors include personality and demographic characteristics; and environmental influences come from social pressure or unique situational characteristics. According to SCT, individuals choose the environments they involved in, and are influenced by the environment incidentally. Furthermore, cognitive and personal factors determine, as well as reshaped by behavior. Finally, behavior in a given situation is affected by environmental or situational characteristics, which are impacted by behavior in turn. Bandura denotes the nature of these relationships as "triadic reciprocity." This study concerns the dynamic interactions in organizational value co-creation context.

13.2.2 Organizational Value Co-creation Behavior

This study concentrates organizational value co-creation from the perspective of organizational citizenship behavior (OCB), including organizational identification, altruistic behavior and knowledge sharing behavior.

According to Organ [20], OCB is defined as work-related behaviors that are discretionary, not related to the formal organizational reword system, and, in aggregate, promote the effective functioning of the organization. Williams and Anderson [25] further divided OCB into two major dimensions: (1) behaviors directed at specific individuals in the organization, such as courtesy and altruism (OCB-I); and (2) behaviors concerned with benefiting the organization as a whole, such as conscientiousness, sportsmanship and civic virtue (OCB-O). This study follows this classification of OCB, and adopts organizational identification as OCB-O, and altruistic and knowledge sharing behavior as OCB-I respectively.

13.2.3 Organizational Climate

Organizational climate refers to the perceptions and feelings of organizational members regarding their work environment [12]. It is multi-dimensional in nature and assumed to influence personal motivation and behavior [15, 16]. For example, Schulte et al. [22] maintain that organizational climate plays an important role in understanding organizational members' attitude, and prove individual's evaluation of the climate is positively related to their attitude. Specifically, the more positive the perception of the organizational climate, the more energetic the collective attitude should be.

Different aspects of organizational climate have been identified to be critical drivers for knowledge sharing, such as top management support [6], employee involvement [5], stimulus to develop new ideas [24], an open and freely expressive climate to keep information flowing [14], and reward systems linked to knowledge sharing [4].

13.2.4 Self-efficacy and Outcome Expectations

In terms of personal factors, SCT emphasizes two sets of expectations as the major cognitive forces guiding behavior. The first set of expectations relates to outcome expectations, and the second set encompasses self-efficacy [1–3]. Outcome expectations refer to the judgment of likely consequence of specific behavior. Outcome expectations may be personal, such as senses of pleasure or happiness in behavior, or team-related, such as improved project performance. Meanwhile,

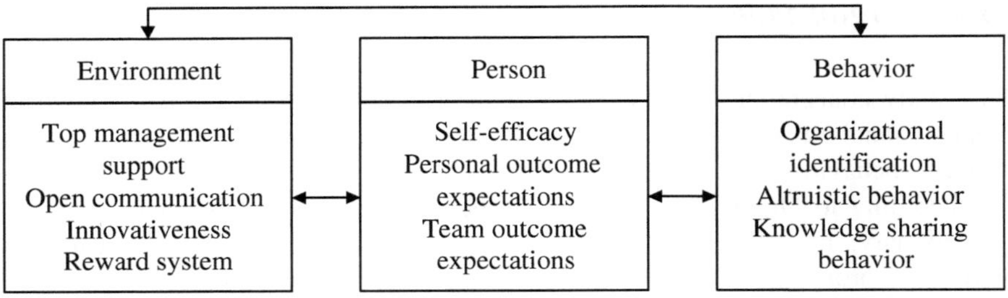

Fig. 13.1 Research framework of this study

self-efficacy is the belief that one possesses the skills and abilities to successfully accomplish a specific task [21].

Though positive outcome expectations are essential for actual behavior, they must be supplemented by positive self-efficacy. People who have higher self-efficacy can be expected to perform related behavior more likely than those with lower self-efficacy. As a result, Bandura [1] posits that self-efficacy is the mastery of behaviors ultimately. In sum, the research framework of this study can be drawn as Fig. 13.1.

13.3 Building a DANP Model for Exploring the Dynamics of Organizational Value Co-creation Behavior

13.3.1 Proposed DANP Model

This study applies DANP [9] to validate the proposed research framework. DANP represents DEMATEL-based ANP that can identify the interdependence among the dimensions and criteria. Major desirable features of these tools are they rely on extensive pair-wise comparisons among system elements; no prior assumptions about their relationships are needed. Then the final influence pattern (directions and degrees) will be objectively derived by mathematic calculations instead of subjectively judgments. In this way, this study can contribute insightful and complementary findings to this research field.

13.3.2 Instruments and Surveyed Experts

In order to explore the dynamic relationships among personal, environmental, and behavioral factors regarding organizational value co-creation behavior, we developed a questionnaire to survey experts who have rich experiences in knowledge-intensive activities. Based on their input, we conduct DANP analysis to identify the dynamic relationships of related constructs. The surveys were conducted in face-to-face manner to make sure experts understand real meanings of research constructs.

We followed suggestions from Northcutt and McCoy's [19, p. 87] and Strauss and Corbin [23] in terms of qualitative and quantitative criteria to recruit 16 domain experts from academic (44 %) and industry fields (56 %) with managerial (23 %) and professional (77 %) and at less three years of work experiences.

13.4 Results and Discussions

After DANP analysis, the influential relationships among dimensions and criteria can be determined and shown as Table 13.1, in which elements are prioritized by their power of influences.

Major results of this study in the context of organizational value co-creation can be described as follows. First, the triadic reciprocity of personal, environmental, and behavioral factors are validated. Since all elements influence and be influenced by all other elements. This conclusion reflects the dynamic and sophisticated real world situation properly.

Second, dominant influencing trends can be identified clearly. For example, from dimension's point of view, environmental factors affect personal factors and behavioral factors; personal factors affect behavioral factors. The implication of this finding is that in order to enhance value co-creation behavior, managers need to improve environmental factors first to motivate personal factors to actually conduct desired behaviors. Similarly, in terms of organizational value co-creation behavior, organizational identification affects altruistic behavior and knowledge sharing behavior; and altruistic behavior affects knowledge behavior. It is evident that organization identification lays the foundations of general altruistic behavior and specific knowledge sharing behavior.

Table 13.1 Dominant influences

Elements	Influences	
	Primary	Secondary
Dimensions	Environmental	Personal, behavioral
Criteria		
Environmental	Management support	Reward system, open communication, innovativeness
Personal	Self-efficacy	Personal outcome expectation, team outcome expectation
Behavioral	Organizational identification	Altruistic behavior, knowledge sharing behavior

These findings can be expected to provide helpful guidance for the effective promotion of organizational value co-creation behavior to leverage the operant resources to their maximum potentiality.

References

1. Bandura A (1982) Self-efficacy mechanism in human agency. Am Psychol 37(2):122–147
2. Bandura A (1986) Social foundations of thought and action: a social cognitive theory. Prentice-Hall, Inc., Englewood Cliffs
3. Bandura A (1997) Self-efficacy: the exercise of control. Freeman, New York
4. Bartol KM, Srivastava A (2002) Encouraging knowledge sharing: the role of organizational reward systems. J Leadersh Organ Stud 9(1):64–76
5. Bock GW, Kim YG (2002) Breaking the myths of rewards: an exploratory study of attitudes about knowledge sharing. Inf Res Manage J (IRMJ) 15(2):14–21
6. Bock GW, Zmud RW, Kim YG, Lee JN (2005) Behavioral intention formation in knowledge sharing: examining the roles of extrinsic motivators, social-psychological forces, and organizational climate. MIS Q 29(1):87–111
7. Bostrom RP, Heinen JS (1977) MIS problems and failures: a sociotechnical perspective part I: the cause. MIS Q 1(3):17–32
8. Cabrera EF, Cabrera A (2005) Fostering knowledge sharing through people management practices. Int J Hum Res Manage 16(5):720–735
9. Chen F-H et al (2011) A balanced scorecard approach to establish a performance evaluation and relationship model for hot spring hotels based on a hybrid MCDM model combining DEMATEL and ANP. Int J Hospitality Manage 30(4):908–932
10. Compeau D, Higgins CA, Huff S (1999) Social cognitive theory and individual reactions to computing technology: a longitudinal study. MIS Q 23(2):145–158
11. Davenport TH, Prusak L (2000) Working knowledge: how organizations manage what they know. Harvard Business Press, Boston
12. Denison DR (1996) What is the difference between organizational culture and organizational climate? A native's point of view on a decade of paradigm wars. Acad Manag Rev 21 (3):619–654
13. Foss NJ, Pedersen T (2002) Transferring knowledge in MNCs: the role of sources of subsidiary knowledge and organizational context. J Int Manag 8(1):49–67
14. Jarvenpaa SL, Staples DS (2000) The use of collaborative electronic media for information sharing: an exploratory study of determinants. J Strateg Inf Syst 9(2):129–154
15. Kaya N, Koc E, Topcu D (2010) An exploratory analysis of the influence of human resource management activities and organizational climate on job satisfaction in Turkish banks. Int J Hum Res Manage 21(11):2031–2051
16. Litwin GH, Stringer Jr RA (1968) Motivation and organizational climate. Harvard Business Press, Boston
17. Lusch RF et al (2007) Competing through service: insights from service-dominant logic. J Retail 83(1):5–18
18. Lusch RF et al (2008) Toward a conceptual foundation for service science: contributions from service-dominant logic. IBM Syst J 47(1):5–14
19. Northcutt N, McCoy D (2004) Interactive qualitative analysis: a systems method for qualitative research. Sage Publications, Inc., Thousand Oaks
20. Organ DW (1988) Organizational citizenship behavior: the good soldier syndrome. Lexington Books/DC Heath and Company, Lexington
21. Ormrod JE (2006) Essentials of educational psychology. Pearson Merrill/Prentice Hall, Upper Saddle River

22. Schulte M, Ostroff C, Shmulyian S, Kinicki A (2009) Organizational climate configurations: relationships to collective attitudes, customer satisfaction, and financial performance. J Appl Psychol 94(3):618–643
23. Strauss A, Corbin J (2007) Basics of qualitative research: techniques and procedures for developing grounded theory. Sage Publications, Inc., Thousand Oaks
24. Taylor WA, Wright GH (2004) Organizational readiness for successful knowledge sharing: challenges for public sector managers. Inf Res Manage J (IRMJ) 17(2):22–37
25. Williams LJ, Anderson SE (1991) Job satisfaction and organizational commitment as predictors of organizational citizenship and in-role behaviors. J Manage 17(3):601–617
26. Wood R, Bandura A (1989) Social cognitive theory of organizational management. Acad Manag Rev 14(3):361–384

Chapter 14
FontCloud: Web Font Service for Personal Handwritten, Ancient, and Unencoded Characters

Jeng-Wei Lin, Feng-Sheng Lin, Yu-Chun Wang, Jan-Ming Ho and Ray-I Chang

Abstract Text is still the most convenient and important media for people to ccommunicate with others. Nowadays, designers can use various font setting to demonstrate their ideas in text. At the same time, new characters are invented to expand the power of text. However, people have to wait before these characters are encoded in computers and related fonts are released. It is not easy for ordinary users to create fonts, especially of Chinese characters. In this paper, a font cloud offering web font service is presented where users can create their personal fonts easily from the images of characters. These characters images could be scanned from ancient calligraphy works or someone's handwritings on papers or electronic pads. After preprocessed, these images are vectorized, unified to their codepoints in Unicode, and packed into a user-specified font stored in font cloud. A JavaScript is automatically generated. Users can embed the JavaScript in their web pages to use the font in font cloud and show the designed text effect. The prototype system shows that the proposed font cloud can help users greatly.

Keywords Web font · FontCloud · Handwriting · Chinese character · Ancient character · Unencoded character

J.-W. Lin
Department of Information Management, Tunghai University, Taichung, Taiwan
e-mail: jwlin@thu.edu.tw

F.-S. Lin · Y.-C. Wang · J.-M. Ho
Institute of Information Science, Academia Sinica, Taipei, Taiwan
e-mail: skyrain@iis.sinica.edu.tw

Y.-C. Wang
e-mail: zxaustin@iis.sinica.edu.tw

J.-M. Ho
e-mail: hoho@iis.sinica.edu.tw

R.-I. Chang (✉)
Department of Engineering Science and Ocean Engineering, National Taiwan University, Taipei, Taiwan
e-mail: rayichang@ntu.edu.tw

J.J. (Jong Hyuk) Park et al. (eds.), *Future Information Technology - II*,
Lecture Notes in Electrical Engineering 329,
DOI 10.1007/978-94-017-9558-6_14

14.1 Introduction

Due to the rapid improvement in computer and network technologies, digital multimedia messages have become more and more popular. However, text is still the most popular and important media for people to communicate with others. Many efforts had been made. Nowadays, people can use many different font setting to show amazing effect of a text. In the market, many fonts have been developed so that designers can choose unique fonts to demonstrate their idea gracefully. Google Fonts [2] provides free web font service with more than 600 font families so that web designers can exhibit a colorful text without worrying whether the specified fonts are installed in client devices. Nevertheless, people are still unsatisfied due to the lack of fonts of newly created interesting and inspiring characters and symbols. For example, Unicode version 6 [4] in 2010 encoded emoji (emotion Hanzj) characters, such as ☹ and ☺, to expand the power of text [11]. It usually takes a very long time, probably several years, for standard organizations to encode these characters in a computer charset encoding. Even though these characters are encoded, users have to wait for the release of related fonts. People sometimes want to render a text in their unique handwritten style. In fact, it is commonly believed that handwritings can reflect one's personality, emotion, feeling, education level, and so on. This is especially true in Chinese calligraphy. However, it is not easy for ordinary users to make a font of their personal handwritings. As a result, people usually have to search a suitable font in the market, which is usually made by professional font designers or companies.

Many tools had been developed for font engineers. Some of them are propriety, and some are free in public domain. These tools usually require a certain level of prior knowledge of computer systems, charset encodings, and font specifications [7]. Users have to master skills of computer graphics and image processing. Figure 14.1 shows some snapshots of FontForge [12], an open-source package for font design. It allows users to create or modify the vector glyphs for characters according to their codepoints in a font file, as shown in Fig. 14.1a. For each character, the glyph creation follows these steps: (1) optionally importing a bitmap image of the target character, (2) tracing the edge of the character image and fine tuning it to make a

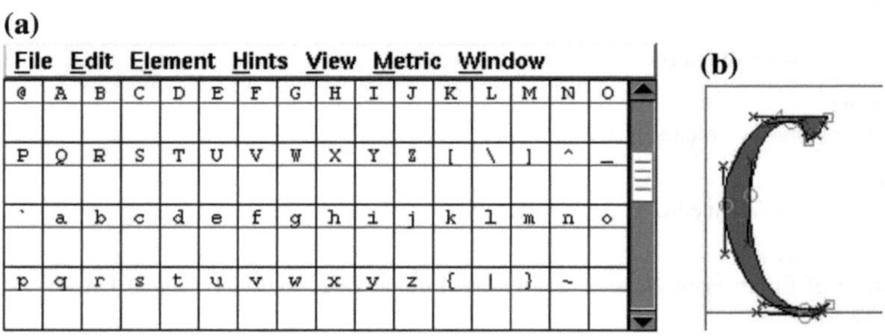

Fig. 14.1 Snapshots of FontForge: (**a**) characters in a font file, (**b**) the vector glyph of font "C"

vector glyph, as shown in Fig. 14.1b. It usually takes a lot of time for font engineers to beautify the glyph in the second step. After all vector glyphs are made, the users can do a final adjustment, such as kerning of characters.

These tools usually do not support Chinese characters well. Furthermore, a Chinese font may contain several thousands of Chinese characters and most of them are much more complicated than English and Latin letters. Many studies have been done for glyph creation of Chinese characters [5, 13–20]. However, it is out of the scope of this paper.

In this study, an on-demand web font service, referred to as font cloud, is presented. We performed a process reengineering in font generation. Rather than to create vector glyphs of all encoded characters in a charset encoding one by one according to their codepoints, people create vector glyphs incrementally in an on-demand manner. A user can handwrite some sentences, phrases, or a simple text comprised of wanted characters and symbols on a paper (or an electronic pad). The images of these handwritten characters are scanned, preprocessed and digitized into vector glyphs, and then unified to their corresponding codepoints. When a set of encoded vector glyphs are made, the user can pack them into a user-specific font file stored in font cloud. After the font is generated, a JavaScript is automatically generated for web font service. Web designers can embed the script in their pages and use the font stored in font cloud.

We note that the user does not have to create vector glyphs for all characters in a charset encoding at a time. If the user finds that some characters display improperly, he or she can redo the aforementioned process to complete the font increasingly.

A prototype system of font cloud has been built. Several preliminary experiments were performed, including works of ancient Chinese calligraphers and several volunteers. The results show our system can help users create personal fonts of ancient, handwritten, or unencoded characters and symbols easily and quickly.

This paper is organized as follows. We describe the system architecture of font cloud and the font creation process in Sect. 14.2. Then, we present the preliminary experiment results on the prototype system in Sect. 14.3. Finally, Sect. 14.4 gives our conclusion and intended future works.

14.2 System Flow and Architecture of Font Cloud

Figure 14.2 shows the system architecture of font cloud. As mentioned, users of our system do not have to create vector glyphs of all characters and symbols in a charset encoding. When a user wants to make a personal font, at the beginning our system will provide an empty base font. The user can give the font a name for reference. Then the user can follow the steps to add vector glyphs of wanted characters into font cloud.

1. The user prepares a digital image of characters or symbols. It could be a scanned image from works of famous Chinese calligraphers or anyone. It could also be

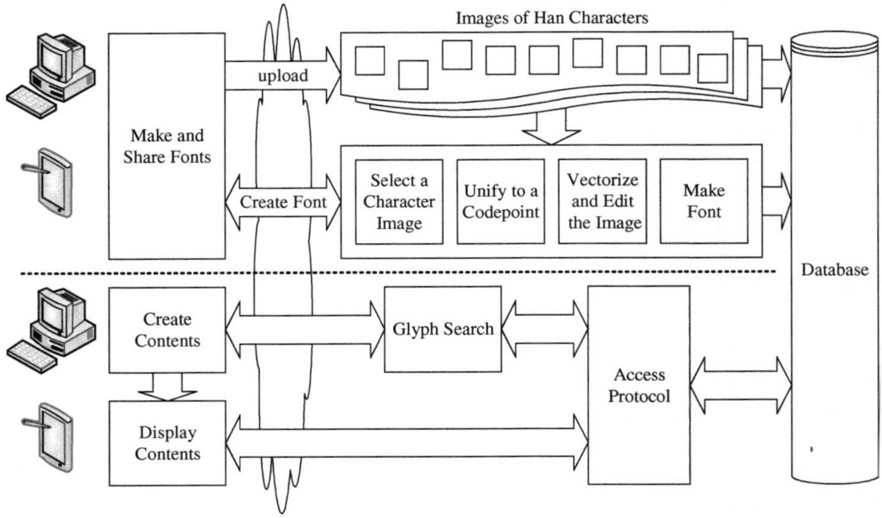

Fig. 14.2 System architecture

captured by a touch panel or electronic drawing pad on which the user can handwrite wanted characters and symbols.

2. The user can optionally apply several image processing techniques to improve the quality of the digital image, such as dithering, erosion, dilation, and so on.
3. The user uploads the image to font cloud, through a user-friendly web interface.
4. The user iteratively adds a vector glyph of a target character into the database.
   (a) The user crops the image of the target character from the uploaded image.
   (b) The user has to unify the cropped image to the codepoint of the target character in a charset encoding used in computers. Currently, Unicode is adopted.
   (c) The cropped image is vectorized to a SVG image [8]. The user can further fine tune the vector glyph by a web-based SVG editor.
   (d) Once the vector glyph is made, the user submits the cropped image, vector glyph, its codepoint, and related information into the database.
5. When a set of encoded vector glyphs are made, the system can pack them into the user-specific font and stored it in font cloud. As well, a JavaScript is automatically generated for web font access.

A user can online access the font via the JavaScript generated by font cloud. Currently, one can use web fonts in HTML5 pages. The user can look up the fonts stored in font cloud and embed the related JavaScript in a page. When an enabled browser is used to access the page, it can use the JavaScript to download the font from font cloud and render the text in the designed font setting, as shown in Fig. 14.3.

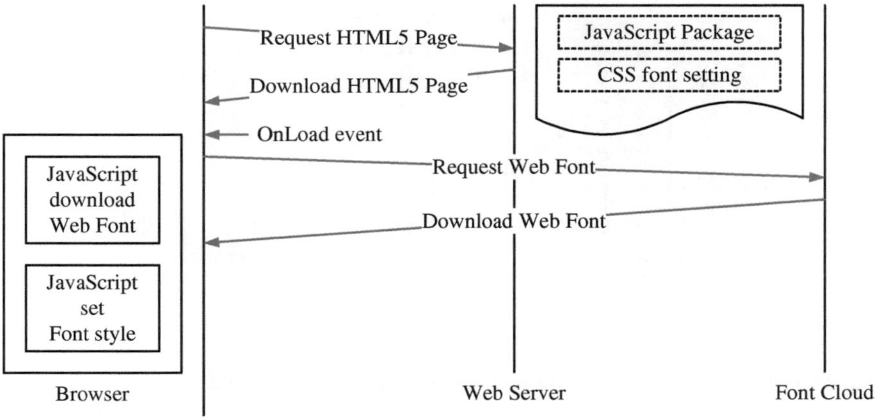

Fig. 14.3 HTML5 web font service of FontCloud

14.3 Prototype of the Font Cloud

We have built a prototype system of font cloud. It integrated many open-source tools to perform aforementioned steps, such as ImageMagick [3] for bitmap image preprocessing, potrace [9] for bitmap image vectorization, SVG-edit [10] for vector image drawing, and FontForge [12] for font manipulation. Figure 14.4 shows some snapshots of the prototype of font cloud.

Figure 14.5 demonstrates a web page that uses the font generated in the prototype system. The work, 日月如馳帖, in the left of Fig. 14.5, was made by a famous ancient Chinese calligrapher 王羲之. The upper-right is rendered by the generated font, and the bottom-right is rendered by a popular font, 細明體. Readers can notice that both the characters 月 and 如 appeared twice and had different

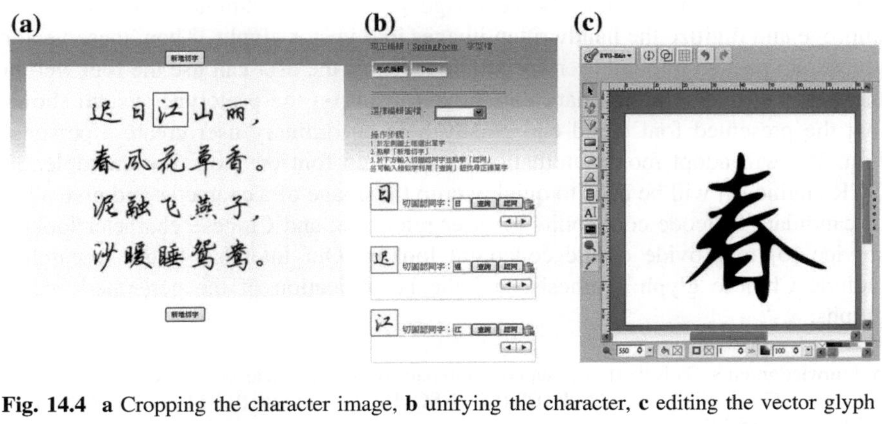

Fig. 14.4 **a** Cropping the character image, **b** unifying the character, **c** editing the vector glyph

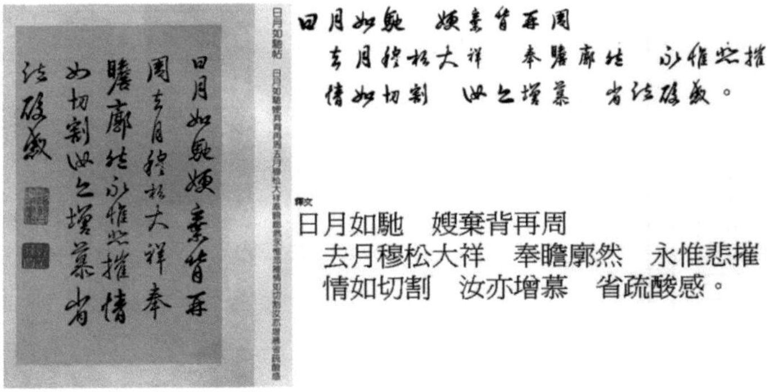

Fig. 14.5 A simple demonstration of our prototype system of font cloud

appearances in the original work. However, since there can be only one vector glyph for each target character in a font, 月 has the same appearance in the upper-right, and so does 如.

The implementation also includes a handwriting module for an electronic drawing pad on which users can handwrite characters and symbols. Thus, when a user is unsatisfied with the glyph of a character, the user can easily create a new one.

14.4 Conclusion and Future Work

In this paper, we present a font cloud offering web font service where users can create their personal fonts easily from the images of wanted characters. Rather than to create vector glyphs of all characters in a charset encoding at one time, an on-demand approach is adopted. A user of font cloud can arbitrarily handwrite some characters or symbols. Some image processing techniques are applied to improve and digitize the handwritten images into vector glyphs. Then, these vector glyphs are packed into the user-specific font. Thus, the user can use the font web in the web pages. The preliminary experiment result on the prototype system shows that the presented font cloud can greatly help an ordinary user create a personal font. We will adopt more automation to speed up font creation. For example, an OCR engine [1] will be used to quickly crop the image of a character and give a list of candidate Unicode codepoints for user reference, and Chinese character lookup service [6] to provide online codepoint lookup. Our intended future researches include Chinese glyph synthesis, and the beautification of the generated vector glyphs.

Acknowledgments This work was supported in part by National Science Council, Taiwan (R.O. C.), under Grant NSC 102-2221-E-029-023, NSC 102-2221-E-001-013-MY3, and NSC 102-2221-E-001-015-MY3.

References

1. Chou C-H, Lin W-H, Chang F (2010) A binarization method with learning-built rules for document images produced by cameras. Pattern Recogn 43(4):1518–1530
2. Google Fonts. https://developers.google.com/fonts/
3. ImageMagick. http://www.imagemagick.org/
4. Juang D-M, Wang J-H, Lai C-Y, Hsieh C-C, Chien L-F, Ho J-M (2005) Resolving the unencoded character problem for chinese digital libraries. In: 5th ACM/IEEE joint conference on digital libraries
5. Lim SB, Kim MS (1995) Oriental character font design by a structured composition of stroke elements. Comput Aided Des 27(3):193–207
6. Lin J-W, Lin F-S (2012) Unicode Han character lookup service based on similar radicals. Int J Smart Home 6(3):99–106
7. Phinney TW (2004) TrueType, PostScript type 1, and OpenType: what's the difference, version 2.36. http://blogs.adobe.com/typblography/TT%20PS%20OpenType.pdf
8. World Wide Web Consortium (2011) Scalable Vector Graphics 1.1, 2nd edn. W3C Recommendation. http://www.w3.org/TR/SVG/
9. Selinger P (2009) Potrace: a polygon-based tracing algorithm. http://potrace.sourceforge.net/
10. SVG-edit. https://code.google.com/p/svg-edit/
11. The Unicode Consortium (2010) The Unicode Standard, version 6.0. http://www.unicode.org/
12. Williams G (2003) Font creation with FontForge. In: EuroTeX 2003, TUGboat, vol 24, pp 531–544
13. Wong PYC, Hsu SC (1995) Design chinese typeface using components. In: Computer software and applications conference
14. Wong HTF, Ip HHS (2000) Virtual brush: a model-based synthesis of chinese calligraphy. Comput Graph 24:99–113
15. Xia W, Jin L (2010) A Kai style calligraphic beautification method for handwriting chinese character. In: International conference on document analysis and recognition
16. Xu S, Lau FCM, Cheung WK, Pan Y (2005) Automatic generation of artistic chinese calligraphy. IEEE Intell Syst 20(3):32–39
17. Zhang J, Mao G, Lin H, Yu J, Zhou C (2001) Outline font generating from images of ancient chinese calligraphy. In: Transactions on edutainment V (Lecture notes in computer science), vol 6530. Springer, Berlin, pp 122–131
18. Zhang J, Yu J, Lin H (2007) Capturing character contours from images of ancient chinese calligraphy. In: 2nd workshop on digital media and its application in museum and heritage
19. Zhu X, Jin L (2008) Calligraphic beautification of handwritten chinese characters: a patternized approach to handwriting transfiguration. In: International conference on frontiers in handwriting recognition, pp 135–140
20. Zitnick C (2013) Handwriting beautification using token means. ACM Trans Graph 32(4):53

Chapter 15
Implementing Globally Unique Identifier Architecture in Date Collection for a Health Management Study in Taiwan Aboriginal Tribe

Jui-Fu Hung, Hsiao-Hsien Rau, Chien-Yeh Hsu, Shih-Chang Chen, Duu-Jian Tsai, Yang Fann, Joshua Park and Joshua Eng

Abstract In present, more and more attention has been put on the effort of protection of personal data and also new encryption technologies are derived. In this article, we introduce a personal information protection technology, global unique identifier (GUID), which is developed by National Institutes of Health U.S. and used for personal health information protection. The GUID is an identifier that allows researchers to associate and share data specific to a study participant without using or exposing personally identifiable information (PII). The GUID is made up of random alphanumeric characters and is not directly generated from PII/Protected Health Information (PHI). It has been approved by the NIH/USA office General

---

J.-F. Hung · H.-H. Rau · C.-Y. Hsu · D.-J. Tsai
Graduate Institute of Biomedical Informatics, Taipei Medical University, Taipei, Taiwan
e-mail: m610100004@tmu.edu.tw

H.-H. Rau
e-mail: d110097003@tmu.edu.tw

C.-Y. Hsu
e-mail: cyhsu@tmu.edu.tw

D.-J. Tsai
e-mail: dj.tsai@msa.hinet.net

S.-C. Chen (✉)
Graduate Institute of Patent, National Taiwan University of Science and Technology, Taipei, Taiwan
e-mail: scchen1980@ntu.edu.tw

Y. Fann · J. Park · J. Eng
Intramural IT and Bioinformatics Program, NINDS, NIH, HHS, Bethesda, USA
e-mail: Fann@ninds.nih.gov

J. Park
e-mail: joshua.a.park@gmail.com

J. Eng
e-mail: jeng@sapient.com

J.J. (Jong Hyuk) Park et al. (eds.), *Future Information Technology - II*,
Lecture Notes in Electrical Engineering 329,
DOI 10.1007/978-94-017-9558-6_15

Counsel for secure sharing of anonymized database. In our study, we used GUID as key identifier for health data collection from aboriginal tribe in a health promotion project.

15.1 Introduction

A globally unique identifier (GUID) is a unique reference number used as an identifier in computer software. The term GUID typically refers to various implementations of the universally unique identifier (UUID) standard [1–3]. In a medical study, we need a unique subject Identifier to associate same subject participated in any multi-site study or used in data repositories (e.g. National Cancer Study, Data warehouse or BioBank). To follow the personal information protection act, subject privacy and confidentiality need be protected and de-identified and anonymized data security need to be ensured [4, 5]. When the study is conducted in multiple sites, we need to enable data sharing and exchange across studies, sites and systems without using any part of personal identifiable information (PII) [6]. In addition, when collecting data from different sites of the same subject to be associated without transmitting or exposing any PII [7, 8].

After entering a set of PII of the subject, such as name, birth date, gender and city at birth, the GUID client will use a one-way hash function to processes these PII and generate several intermediary codes, and transmits the codes to GUID server under a secure manner. The application on the server will compares the transmitted hash codes with the existed GUID database. If there is no match, the server application generates a new GUID randomly according to a pre-defined format, and stores the association between the hash codes and the GUID in the repository. If the codes match those from a previous transmission, the associated GUID is obtained. The researcher obtains the GUID from the server application. The association between the GUID and the participant's research datasets is established and stored locally. For the purpose of sharing the participant's clinical data with other researchers or sending to a data repository, the researcher removes all identifying information except the GUID. Once the de-identified dataset is transmitted to the repository, the dataset becomes anonymized and can be linked with other datasets that follow the same regulation.

15.1.1 Function and Workflow of GUID System

The GUID service functions are composed and described as the follows:

GET This function uses the hash function to generate a unique valid GUID(s). It uses the available minimum required PII information provided in the system.

GETBATCH This function allows multiple users GUID requests by a single call to the server. The GUID(s) request numbers are no limit. Depends on the computing power and the network, generally the batch function can processed 50 users at a same time with a 5 min delay between each 50 users processed.

GETINVALIDGUID Generates a valid GUID and a pseudo-GUID that format is different. PII is not required to get Pseudo-GUID and pseudo-GUID is not based on PII to be a unique ID. This function can only be used when the users do not have enough PII and therefore cannot use the GET function.

TESTCONNECTION This function is to test the connection of the GUID server. It will not store any information data to the GUID database.

DOESGUIDEXIST This function can let users to check if the GUID(s) is already existed in the database.

CONVERTPSEUDOGUID This function allows users convert pseudo-GUID(s) to valid GUID(s). But the users need to provide pseudo-GUID(s) and minimum required PII fields.

15.1.2 The Security of GUID System

The security at application and network/server levels of GUID is provided to reduce the potential of the algorithm to be analyzed and decompiled at the client application. The subnet mask of study site IP address is restricted. Encryption is between client and server connection and user authentication. To get a de-identified GUID, hackers attack will need all the combinations of binary data fields with hash over inverse probability of codes. But, if the hacker is the site manager or original user who already access to PIIs, the GUID of patient cannot be re-identified. The PIIs of any patient are never stored at GUID server and transmit offsite.

15.2 Methods

15.2.1 PII Choice, Arrangement and Combination

Depends on different purpose of every research, which PII should be choice is a very important issue. Basically, it is very associated with the repetition rate and uniqueness of every PII fields. The PII fields should be easy to obtain e.g. gender, birth data, country of birth etc [9, 10]. Any study project or repository for combination of required data fields should not have inverse probability exceeding the number to protect any potential duplicates. The arrangement and combination of key in PII fields determines the probability of matching the same GUID from a set of three, five or more hash codes. A PII which does not change over the life time and

Table 15.1 Usually used PII fields

PII field name	Abbreviation	Inverse probability	Require
Government issued ID or national ID	GIID	100,000,000	No
Subject's complete legal given name at birth	FN	200	Yes
Subject's complete legal family name at birth	LN	3,000	Yes
Day of month of birth	DOB	30	Yes
Month of birth	MOB	12	Yes
Year of birth	YOB	100	Yes
Physical sex of subject at birth (M/F)	SEX	2	Yes
Name of subject's born city/municipality	COB	300	Yes
Country of birth	CNOB	300	Yes
Government-issued of national ID	COUNTRYID	300	Yes

Table 15.2 Use PII fields to form hash codes

Hash code	1	2	3	4	5	6	Probability
1	YOB	DOB	GIID	Sex			6E+12
2	FN	LN	COB	DOB	MOB		6E+11
3	FN	YOB	MOB	COB	CNOB		2E+10
4	FN	LN	COB	Sex	YOB	CNOB	6E+12

uniquely specified is subjected to be used to generate unique hash code. The PII fields that are usually used and their inverse probability are shown in Table 15.1.

This Study use Hash Algorithm to generate 4 sets of Hash Code, The combination of PII fields and the inverse probability is listed in Table 15.2. PIIs are stored in local and the all data collected from any subject are associated with a unique GUID.

Hash Codes is a cryptographic function. There have four main properties of ideal cryptographic hash function:

1. Easy to compute hash value for any given string.
2. Infeasible to find a string that has a given hash.
3. Infeasible to modify a string without changing the hash.
4. Infeasible to find two different string with the same hash [11, 12].

The one way hash codes are transmitted to the center data repository. But with the one way hash codes and data repository are not revealing any PII. All of PII fields are no spaces, no punctuation, only uppercase letters and numbers to keep the data consistency. The combination of PII fields is concatenated to the pattern. Using a one way hash algorithm with a private key for a hash code is converted by all of combination of PII fields. To indicate the number of missing values for that code, it will add an additional byte to add to each resulting code. The numbers of hash codes are checked by complexity and availability of data fields as the inverse probability. To ensure a perfect match at least one hash code should contain all of required data filed. The combination of PII fields was chosen such that each combined inverse probability of each hash code is sufficiently high to confidently delineate subjects.

15.3 Result and Discussion

15.3.1 The GUID Application Use in Aboriginal Tribe Health Care

This Study applicate GUID in Date Collection for a Health Management Study in Taiwan Aboriginal Tribe. Data are collected from different local clinical service sites. When a new subject is entering the system, the GUID server will generate a new GUID for a new subject and the clinical data with the GUID will be transmitted to and store in a central database. The PII will not be used as an identifier for data exchange and research purpose. If the same person goes to more than one local clinical service site, the same GUID will be applied to the person. The GUID can be used to retrieve a patient's data from the local clinical service site. A clinical data repository for our research combining different data from multiple local clinical service sites. The architecture of our system and data flow are shown in Fig. 15.1.

The data we collected include diagnosis (include: hypertension, diabetes, Asthma etc.), medicine use history, life history (include: alcohol use, tobacco use, exercise, food nutrition etc.), laboratory examination history [include: Glycated Hemoglobin (HbA1c), total cholesterol (TCH), High Density Lipoprotein Cholesterol (HDL-C), Low Density Lipoprotein Cholesterol (LDL-C), Uric Acid (UA)].

15.3.2 Benefit of the GUID

In order to make the identity information of the patients are not exposed, developing GUID to be used in different clinical sites and research teams will give each identity information of a patient to get a unique ID. And it needs the patient's basic information to be the PII fields. The fields include last name, first name, gender, birth data and country of birth. We used a hash table to generate a unique GUID. The benefit is all of these PII fields are from the sample original data, and all of these are easy to remember. Generated GUID is hard to duplicate. And it does not have the national personal ID, so it is even safer to use. And even if the patient forgot his/her GUID, just offer his/her information, the GUID can be regenerate.

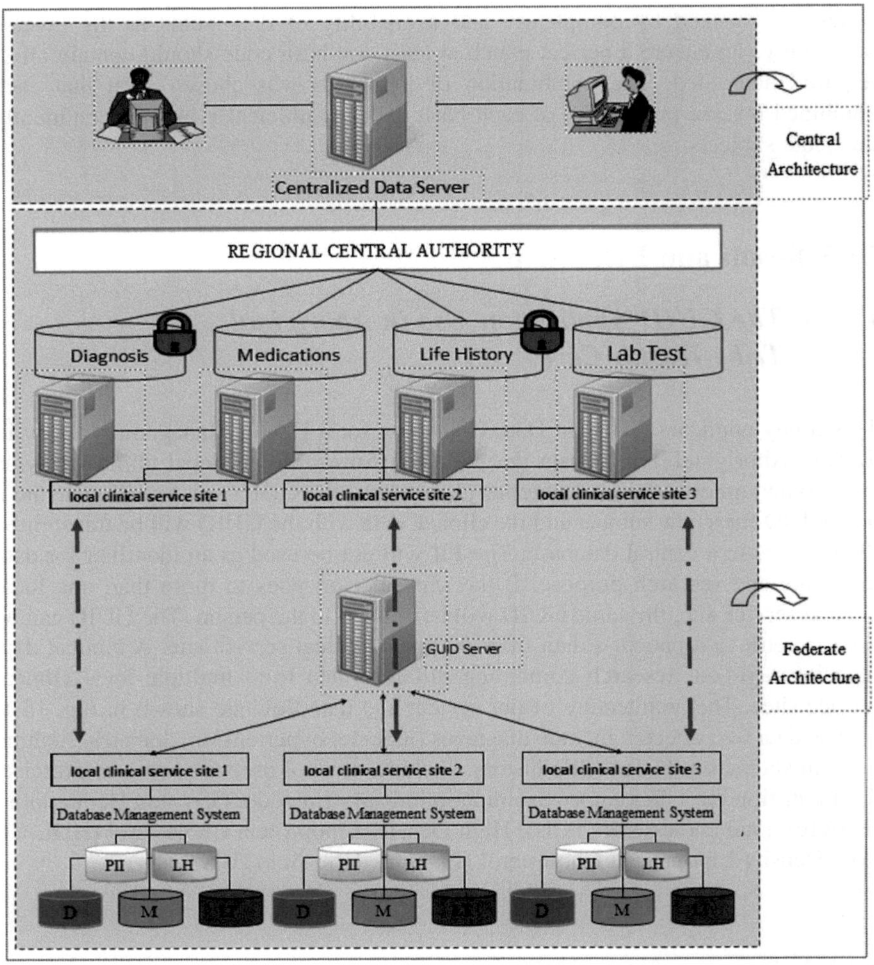

Fig. 15.1 GUID applicate architecture

15.3.3 Effectiveness Evaluation

It is always difficult to collect data from aboriginal tribe because of data security and human right protection. One of the benefit of using the GUID system is that it will let the personal health information to be value-added to produce a high quality research data without exposing personally identifiable information, since the PIIs will be left in the local clinical service sites and not transmitted to other places. It will promote the study of public health decision-making. For example, the integration of data standardization for a specific disease information platform, and data analysis dementia information systems. By using the GUID system, de-identified patient's data is stored in database and that can be effectively accessed and analyzed without violating the privacy of the patients.

Acknowledgments This study was supported and received funding from "Planning Smart Living Labs for Aging in Place Project", which is sponsored by National Science Council in Taiwan (102-2627-E-038-001).

References

1. Boneh D (1999) Twenty years of attacks on the RSA cryptosystem. Not Am Math Soc 46 (2):203–213
2. BS 10012 (2009) Data protection—specification for a personal information management system, May 2009
3. http://law.moj.gov.tw/eng/LawClass/LawAll.aspx?PCode=I0050021
4. http://en.wikipedia.org/wiki/Data_Encryption_Standard
5. http://www.differencebetween.net/technology/difference-between-des-and-aes/#ixzz2rJXYVgbS
6. http://en.wikipedia.org/wiki/RSA_(cryptosystem)
7. https://www.math.sunysb.edu/~scott/Book331/RSA_Public_key.html
8. ISO 29100 Privacy Framework, Rannenberg K (WG5 Convener), Sténuit C (Co-editor 24760), Yamada A (Editor 24761), Weiss AS (Editor 29100/29101) (2007) Working group 5 identity management and privacy technologies within ISO/IEC JTC 1/SC 27—IT security techniques (presentation), 30 Sept 2007
9. ISO (2008) Health informatics—information security management in health using ISO/IEC 27002:2008-07-01
10. Organization for Economic Cooperation and Development (OECD) (2013) OECD guidelines on the protection of privacy and transborder flows of personal data, July 2013. http://www.oecd.org/internet/ieconomy/oecdguidelinesontheprotectionofprivacyandtransborderflowsofpersonaldata.htm
11. Kuo Y-H (2009) The research of power analysis against AES. Unpublished master's thesis, National Central University, Taiwan, June 2004
12. Chen J-Z (2002) The research of Rijndael against physical cryptanalyses. Unpublished master's thesis, National Central University, Sept 2009

Chapter 16
Language Learning in Cloud: Modular Role Player Game-Distance-Learning System Based on Voice Recognition

Ming-Shen Jian, Jun-Hong Shen, Tien-Chi Huang, Yu-Chih Chen and Jun-Lin Chen

Abstract In this paper, the modular role player game-distance-learning system based on voice recognition for language learning in cloud is proposed. Each designed modular game content can be embedded into the learning game platform in cloud. By connecting to the external recognition resource, the pronouncing of learners can be recognized. Only when the pronouncing of the learner can be recognized as the keyword, the further game will be started. By designing different scenarios, different vocabularies or conversations can be added into the RPG game. The information about users in the individual game learning system can be stored in the cloud database. Then users can connect to the learning-game system via the Internet for continuous learning. By using the cloud platform, the users can connect to learning system anytime and anywhere. Furthermore, the students can learn advanced courses themselves by playing the advanced scenarios of the game-learning system without the assistance of a teacher.

Keywords Cloud computing · Game-learning · Language learning · RPG

This research was supported in part by Asia University under grant no. 102-asia-51.

M.-S. Jian · Y.-C. Chen · J.-L. Chen
Department of Computer Science and Information Engineering, National Formosa University, Huwei, Yunlin County, Taiwan
e-mail: jianms@nfu.edu.tw

J.-H. Shen (✉)
Department of Information Communication, Asia University, Taichung, Taiwan
e-mail: shenjh@asia.edu.tw

T.-C. Huang
Department of Information Management, National Taichung University of Science and Technology, Taichung, Taiwan
e-mail: tchuang@nutc.edu.tw

J.J. (Jong Hyuk) Park et al. (eds.), *Future Information Technology - II*,
Lecture Notes in Electrical Engineering 329,
DOI 10.1007/978-94-017-9558-6_16

16.1 Introduction

Today, considering the interest of students in learning, different entertainment and multimedia are integrated into the learning content. Randel [7] hypothesizes that using digital games can enhance learning motivation [8]. Compared to traditional classroom teaching and learning, learning via games or digital multimedia can enhance and promote the interest in primary education.

Since players are interested in playing games, they may develop their abilities to solve the problems or stages of the mission designed in a game scenario [7]. In other words, the players will have the motivation to find and learn the key method (s) to solve and complete the mission [1, 2].

However, most of the learning games today provide Question and Answer (Q&A) types of learning for the players. Based on Q&A, the learning games tend to be strict and inflexible with the teaching content. The system player may lose their interest easily. Furthermore, since the method of learning is similar to the original textbook learning, the performance of learning cannot be obviously improved. Since the player may have interest in learning the method for solving the mission designed in the game, self-advanced learning can be achieved. To enhance self advanced learning according to the guidance learning of advance courses, the education theory should be the basic concept of the designed learning game.

In addition, the method of using the game-learning system should be considered. For a student, to install and configure the game-distance-learning system is too difficult. Some limitations of the hardware may also cause a failure in the system installation and configuration. Therefore, to reduce the complexity of game installation and playing, the proposed game-learning system should be established on a remote game server without players' manual configuration. Since the total amount of system users changes all the time, a dynamic number of servers provided for users can be the solution. Therefore, virtual machines (VMs) on cloud computing that are dynamically established in the cloud are used in our proposed game-learning system.

In this paper, the modular role player game-distance-learning system based on voice recognition for language learning in cloud is proposed. The proposed design scheme includes a modular creation procedure based on the education theory. Each section and step of the education course can be designed as a game branch or scenario of the entire learning game.

The main contributions of our proposed system are as follows:

1. Based on external voice recognition, the pronouncing of each game-learning user can be individually identified. The pronouncing can be recognized according to the popular recognition engine such as Google voice searching application [4]. The learning game system can integrate the external voice recognition resource without extra workload.
2. Based on cloud, different games corresponding to specific education sections are developed according to the proposed creation procedure and can be integrated into a single larger game platform. In other words, after each individual playing

or learning session, the designed game can be initialized for the next player. Therefore, the expenses for maintaining the learning game can be reduced.

3. The information and playing progress can be recorded individually. In other words, the individual continuous assessment of a game-learning player can be captured. In addition, since the learning progress is digital and can be stored in a database, the teacher or director can observe the ability and potential of an individual student.
4. The learning game is easy to use. In addition, according to the design and the principle of vertical curricular organization, the interest level of each player can be enhanced and maintained. The player can change and be influenced unobtrusively and imperceptibly into further learning.

16.2 The Development of Modular Role Player Game-Distance-Learning System

Since game-learning has many advantages, in order to adaptively design the game-learning system, the mission or solution of each designed game branch or scenario should correspond to the objective of the educational content or conversation scenario. In this paper, a mission, game branch, or scenario of the game is also called an *Act*. Each *Act* can be a conversation or a vocabulary for language learning. After creating an *Act*, each or different *Acts* can be individually played as a small e-learning game or integrated into a larger game with other *Acts*. Figure 16.1 presents the concept of the modular scenario design procedure for game-learning. To create the *Acts*, the designer has to do the following:

1. Develop the scenario or drama of the game unit. According to the design tool, several scenes or events of the entire scenario called *Act*s can be connected or integrated.
2. The designer can separate the drama into several scenes (partitions of the scenario). Each *Act* consists of characters, music, and mission events (conversation topic) which are determined according to different conditions and backgrounds.
3. The design of mission events or conversation topic of the scenario follows the demand or achievement of the teaching content. Especially, the on demand defined answer for the conversation teaching should be clear and unique as possible.
4. Each scenario should be tested to confirm that all the material and events match the conversation topic or achievements of the teaching content. In other words, the related information corresponding to the conversation should be clear enough for voice recognition matching.
5. Several different scenes can be then integrated into one game unit. It means that the original language learning content can be divided into several sub-partitions for partial conversation or vocabulary learning.

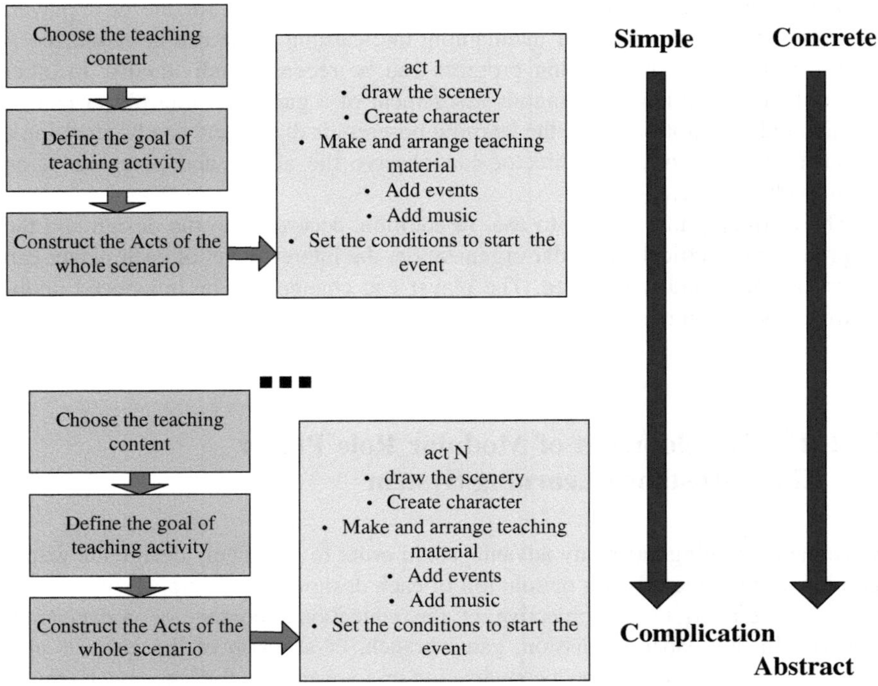

Fig. 16.1 The standardized and modular e-learning game *Act* creation procedure based on vertical curriculum organization principles

Based on the vertical curriculum organization and game playing process, each partial game or *Act* should correspond to the principles which include: learning (1) from simple to complex; (2) from familiar with to not familiar with; (3) from concrete to abstract; (4) from whole to part; (5) from the past to now; (6) pre-condition study first; and (7) concept association method. Thus, players could progress through the game step by step without frustration and depression. Furthermore, the potential of an individual player or learner can be enhanced.

When a player starts the proposed learning game, the complication of the game should be manually selected first. Then, considering the level or characteristics of the player, only the suitable games or conversation stories can be triggered. The e-learning game evaluates the player according to the past records. If the on demand given conditions of the mission (or *Act*) completion cannot be satisfied, it means that the player fails to finish or complete the current mission. In other words, the learner in the current language learning course (or *Act*) fails. The learner should spend more time for the current language learning content. To evaluate the learner, the voice recognition method is involved. According to the recognition result, the learner can be evaluated. If the teaching objects are satisfied, there will be a reward for the player such as a character level upgrade, obtaining new or finite items, etc.

At the same time, the current information and continuous assessment can be updated for that player.

Teaching events will be given to the student step by step according to the following education theory: (1) Give rise to students' motivation. (2) Then, the student will be given a goal of this course similar to the on demand mission completion conditions of the game. (3) A student will be trained, taught, and examined. (4) Only when the student passes the examination which is similar to the player satisfying the mission conditions of the game, the student can be identified as one who possesses the knowledge of this course. Then, students could keep learning this way again and again.

16.3 Integration of the Game-Distance-Learning System, the Cloud System, and Voice Recognition

Since the cloud computing platforms today are popular [3, 5, 9], the Internet-based services can be used for remote client users via any devices. By using the web browser, the cloud users can login the virtual machine for further operation. In addition, every time the cloud user disconnects the virtual machines, the resource for computing will be returned to the resource pool for the repeat using. Furthermore, by configuring the image file of the virtual machine, the virtual machine can be initialized and repeatedly used. Therefore, the learning system can be deployed on cloud for multiple users and for repeat using. In addition to the main learning system, the information related to learners can be also recorded in the database on cloud. In other words, the learning system users only need the devices which have the Internet connection to the cloud without further configuration. It also means that the learning characteristics and parameters of the user recorded in the remote database can be transmitted between the learning game and the third party applications used in school. In other words, via the Internet, the system user can obtain the game-distance-learning service anytime and anywhere.

In addition to the main learning system, to recognize the conversation or vocabulary, especially in listening and speaking, needs the voice recognition. Today, many websites or applications provide the online voice recognition services [4, 6]. In other words, based on the online voice recognition engine, the learning system on cloud provides the API for connection with the online voice recognition engine of the third party. The pronouncing of the vocabulary or conversation can be recognized by using the public voice recognition engine. To increase the accuracy of recognition, further information related to the teaching content or corresponding to the conversation topic will be added for the public voice recognition engine. According to the background information from the learning system, the voice recognition engine can response the results with higher accuracy. Then, according to the voice recognition result, the learning system can match the recognition result and the on demand given answer in the learning game. Only when the recognition

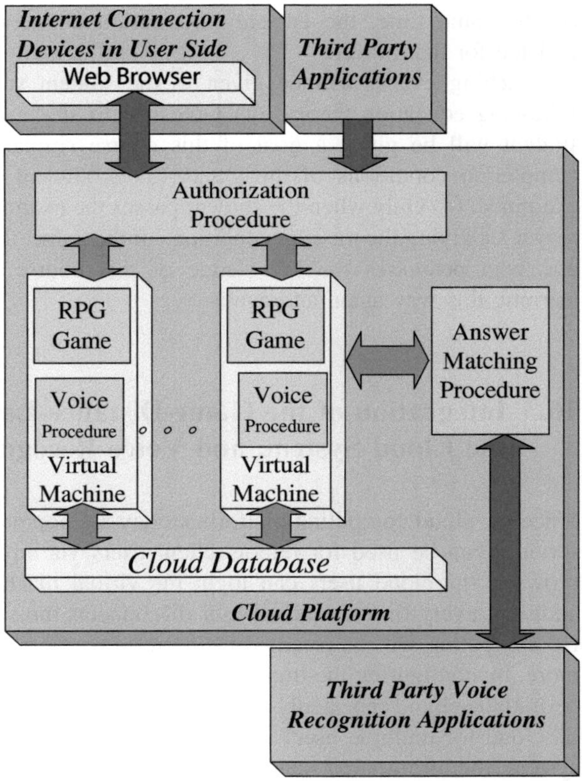

Fig. 16.2 The structure of the proposed system

result can match the answer in the learning game, the learner can be treated as the user who can satisfy the requirement of the conversation or learning content. The system user now can trigger further learning scenario for himself. The structure of the whole proposed system is shown in Fig. 16.2. System users use the devices with Internet connection for interaction with the virtual machine on cloud. To enhance the proposed system, the third party applications such as the student grade management application can also connect to the proposed learning system for learning state querying. All the remote connection should be authorized as the valid connection by using authorization procedure.

16.4 Verification and Implementation Results

In this paper, a game design tool, RPG maker application, is selected to design the game which represents the teaching course for language learning. In this experiment, only when a student's properties or characteristics fit with the on demand condition, the game mission with the next higher level, which indicates more complexity or more multiple concepts, is triggered. In other words, the properties of an individual student must meet the lowest requirement of the new *Act* (game mission). Thus, a student (game player) could play and learn the game step by step.

In the implemented game of the proposed system, the student will play a virtual role for the game mission processing. Only the appearance of each virtual role between different users (player) in each RPG game is the same. Different game missions are playable corresponding to the abilities and characteristics of an individual user.

16.5 Conclusion

In this paper, the modular role player game-distance-learning system based on voice recognition for language learning in cloud is proposed to provide a learning game based on the education theory and vertical curricular organization. Each game-learning user can be identified. The information and playing progress can be recorded individually in the database. The learning game is easy to use. Even a primary student with finite IT knowledge can play the learning game. The experimental results also show that students actually learn the teaching content from the designed game and have obvious progress in the test.

References

1. Bowmaw RF (1982) A Pac-man theory of motivation: tactical implications for classroom instruction. Educ Technol 22(9):14–17
2. Braccy GW (1992) The bright facture of integrated learning system. Educ Technol 32(9):60–62
3. Christensen JH (2009) Using RESTful web-services and cloud computing to create next generation mobile applications. In: 24th ACM SIGPLAN conference companion on object oriented programming systems languages and applications, pp 627–633
4. Google Voice Recognition. https://support.google.com/chrome/answer/1407892?hl=zh-Hant
5. Jian MS, Jhan FJ, Chang HC, Chou TY, Shen JH (2013) Cloud based evolution algorithm with feedback control for emergency logistic. Appl Mech Matter 284–287:3370–3374
6. Microsoft Voice Recognition. http://support.microsoft.com/kb/306537/zh-tw
7. Prensky M (2001) Digital game-based learning. McGraw-Hill, New York
8. Randel JM, Morris BA, Wetzel CD, Whitehill BV (1992) The effectiveness of games for educational purposes: a review of recent research. Simul Gaming 23(3):261–276
9. Rellermeyer JS, Duller M, Alonso G (2009) Engineering the cloud from software modules. In: 2009 ICSE workshop on software engineering challenges of cloud computing, pp 32–37

Chapter 17
Low Complexity Multiuser Interference Cancellation for Frequency Offsets Compensation in Localized OFDMA Uplink Systems

Yu-Jun Won, Jae-Won Suh and Bo-Seok Seo

Abstract In orthogonal frequency-division multiple access (OFDMA) system which allocates subcarriers to different users has much attention because of the advantages in OFDM. However, a well-known problem of OFDM-based system is the weakness for carrier frequency offset (CFO) that leads to a loss of orthogonality among the subcarriers. Furthermore, in OFDMA system, this problem becomes more serious in uplink channel because users have different CFOs. These multiple CFOs (MCFO) bring the multiuser interference (MUI) which is incoming the interference signals from other users by breaking the orthogonality among the users. In this paper, we propose a MUI cancellation scheme for OFDMA uplink systems. In the scheme, two steps of compensation and cancellation in time and frequency domains, respectively, are applied. In the frequency domain, we only consider the interference from both adjacent sides of the band to reduce the computation complexity and parallel interference cancellation (PIC) is employed. Simulation results demonstrate that the proposed method reveals improved performance.

Keywords OFDMA · Multiple carrier frequency offsets · Multiuser interference cancellation · Parallel interference cancellation

---

This work was supported by the National Research Foundation of Korean (NRF-20110014409).

Y.-J. Won · J.-W. Suh · B.-S. Seo (✉)
Department of Electronics Engineering, Chungbuk National University,
Cheongju, South Korea
e-mail: boseok@cbnu.ac.kr

Y.-J. Won
e-mail: heracleo@cbnu.ac.kr

J.-W. Suh
e-mail: sjwon@cbnu.ac.kr

J.J. (Jong Hyuk) Park et al. (eds.), *Future Information Technology - II*,
Lecture Notes in Electrical Engineering 329,
DOI 10.1007/978-94-017-9558-6_17

17.1 Introduction

Orthogonal frequency division multiplexing (OFDM) has many advantages such as high spectral efficiency and ability of eliminating the effects of multipath propagations [1]. Because of its advantages, OFDM is used in several wireless communication or broadcasting systems like as digital audio broadcasting (DAB) [3], terrestrial digital video broadcasting (DVB) [4], and IEEE 802.11a wireless local area network (WLAN) systems [6]. The performance of OFDM depends mainly on how well the orthogonality among the subcarriers is kept at the receiver. Typically, the carrier frequency offset (CFO) which results from the frequency mismatches of oscillators between transmitter and receiver and/or from Doppler effects breaks the orthogonality among the subcarriers. To overcome this problem, much researches have been devoted up to the present [5, 8].

Orthogonal frequency division multiple access (OFDMA) is a multiuser version of the OFDM modulation scheme and has attracted much attention because of the advantages of OFDM. In OFDMA, all the orthogonal subcarriers are divided into a number of mutually exclusive subsets and these subsets are assigned to different users. OFDMA is also as weak to the CFO as OFDM. Especially, in uplink OFDMA systems where several users access to a base station simultaneously, multiple carrier frequency offsets (MCFO) may appear in the received signals of all users. These MCFO bring multiuser interference (MUI) which is the incoming interference from other users' signals.

There have been some literatures to solve the problem of the MUI in uplink OFDMA systems [2, 7]. Choi-Lee-Jung-Lee (CLJL) proposed a post-DFT processing technique by utilizing the basic fact that the multiplication in time domain is equivalent to the circular convolution in frequency domain [2]. However, the CLJL scheme does not reduce the residual MUI and it brings correspondingly performance degradation. Manohar and Sreedhar proposed the method which cancels the MUI in frequency domain by using weighted linear parallel interference cancellation (PIC) after CFO compensation in time domain [7]. The weighted linear PIC is the modified version of the linear PIC, which scales the estimated MUI by weights before cancellation. In the Manohar-Sreedhar scheme, the closed-form expression is needed to calculate the optimum weights. And all the adjacent subcarriers are used to cancel the MUI in a subcarrier symbol resulting in much computational complexity at the receiver.

In this paper, we propose an MUI cancellation scheme for OFDMA uplink systems with multiple frequency offsets. For each user signal, partial PIC is applied in frequency domain, where a subset of the subcarriers is used for MUI cancellation, following the compensation of a CFO in time domain.

The rest of the paper is organized as follows. In Sect. 17.2, we present the uplink OFDMA system model. The proposed MUI cancellation scheme is described in Sect. 17.3. Section 17.4 provides simulation results to verify the performance and conclusions are given in Sect. 17.5.

17.2 Signal Model

We consider an OFDMA uplink system with Q users. We assume that timing synchronization is perfectly achieved at the receiver. We further assume that there are N subcarriers for an OFDM symbol, and M (=N/Q) subcarriers are allocated to each user.

At the transmitter, each user's data are allocated to a designated group of M subcarriers. In this paper, we consider the localized mapping method for the allocation of subcarriers to each user. Without loss of generality, we consider only one OFDM block. After inverse discrete Fourier transform (IDFT), the nth time domain signal of user q is represented as

$$x_n^q = \frac{1}{N}\sum_{k\in S_q} X_k^q \exp\left\{j\frac{2\pi nk}{N}\right\}, \quad 0\le n\le N-1 \tag{17.1}$$

where N is the size of IDFT, X_k^q is a complex symbol allocated to kth subcarrier, and S_q is the set of subcarriers assigned to user q and have the following relations:

$$\bigcup_{q=1}^{Q} S_q = \{0, 1, \ldots, N-1\} \quad \text{and} \quad S_i \cap S_j = \phi \quad \text{for } i \neq j \tag{17.2}$$

Each user's signal is transmitted through an independent multipath channel after guard interval insertion. The channel output for user q is given by

$$y_n^q = x_n^q * h_n^q + w_n^q \tag{17.3}$$

where $*$ denotes convolution operation, h_n^q represents the impulse response of the channel for user q, and w_n^q is the additive white Gaussian noise (AWGN). The received signal is the sum of all Q signals for Q users. Without timing and carrier frequency offsets, the time domain received signal is represented as

$$y_n = \sum_{q=1}^{Q} y_n^q = \frac{1}{N}\sum_{q=1}^{Q}\sum_{k=S_q} H_k^q X_k^q \exp\left\{j\frac{2\pi nk}{N}\right\} + w_n \tag{17.4}$$

where H_k^q means the frequency response of the channel for kth subcarrier of user q, and w_n is AWGN with zero mean and variance σ^2. After discrete Fourier transform (DFT) with guard interval removed, the kth subcarrier symbol of the received signal is given by

$$Y_k = \sum_{n=0}^{N-1} y_n \exp\left\{-j\frac{2\pi nk}{N}\right\}, \quad 0\le k\le N-1. \tag{17.5}$$

With MCFO, the received baseband signal can be represented as

$$r_n = \frac{1}{N}\sum_{q=1}^{Q}\sum_{k\in S_q} X_k^q H_k^q \exp\left\{j\frac{2\pi n(k+\varepsilon_q)}{N}\right\} + w_n \tag{17.6}$$

where ε_q, $q = 1,\ldots, K$, is a normalized CFO of the signal of user q, which is the ratio of the practical carrier frequency offset Δf_q to the subcarrier spacing $1/T_s$ like as

$$\varepsilon_q = \Delta f_q T_s \tag{17.7}$$

where T_s means useful duration of an OFDM block. By substituting (17.6) for y_n of (17.5), we can find the frequency domain signal given by

$$\begin{aligned} R_k^q &= H_k^q X_k^q \frac{\sin \pi\varepsilon_q}{N\sin(\pi\varepsilon_q/N)} \exp\left\{j\frac{\pi\varepsilon_q(N-1)}{N}\right\} \\ &+ \sum_{\substack{l\in S_q \\ l\neq k}} \alpha_{k,l}^{q,q} H_l^q X_l^q + \sum_{\substack{p=1 \\ p\neq q}}^{Q} \sum_{l\in S_p} \alpha_{k,l}^{q,p} H_l^p X_l^p + W_k \end{aligned} \tag{17.8}$$

The first term on the right-hand side represents the effect of the CFO on the desired kth subcarrier symbol resulting in magnitude attenuation and phase rotation of the symbol. The second term means the self-intercarrier-interference (ICI) introduced from adjacent subcarriers of user q itself. The third term is the MUI into the kth subcarrier that comes from the lth subcarrier of the user p. The last term is the DFT of the received AWGN. The $\alpha_{k,l}^{q,p}$ may be regarded as an interference coefficient which is related to the interference to the kth subcarrier of user q from the lth subcarrier of user p by the CFO and expressed as [10]

$$\alpha_{k,l}^{q,p} = \frac{\sin \pi(l-k+\varepsilon_p)}{N\sin\left\{\frac{\pi(l-k+\varepsilon_p)}{N}\right\}} \exp\left\{j\frac{\pi\varepsilon_p(N-1)}{N}\right\} \exp\left\{j\frac{\pi(l-k)}{N}\right\} \tag{17.9}$$

This is a constant determined by the constant N, the subcarrier distance $(l-k)$ and CFO ε_p.

17.3 Proposed MUI Cancellation Scheme

In this paper, in order to eliminate the MUI which comes from the frequency offsets for different users, we assume that all the frequency offsets are known and smaller than the subcarrier spacing (this means that the coarse carrier frequency offset correction is done already [9]).

The proposed method proceeds in two steps: first CFO compensation in time domain and residual MUI cancellation in frequency domain.

17.3.1 CFO Compensation in Time Domain

To compensate the CFO and extract the desired signal for the user q from the received signal, CFO compensation is carried out in time domain by multiplying the received signal with the reverse rotation factor for the user as follows:

$$y_n^q = r_n \exp\left\{-j\frac{2\pi n \varepsilon_q}{N}\right\} \tag{17.10}$$

Then, the DFT of y_n^q is given by

$$Y_k^q = H_k^q X_k^q + \sum_{\substack{p=1 \\ p \neq q}}^{Q} \sum_{l \in S_l} \beta_{k,l}^{q,p} H_l^p X_l^p + W_k' \tag{17.11}$$

Compared with (17.8), we can see that the degradation of the kth subcarrier symbol except for channel effect is compensated, the self-ICI components are removed, and the MUI components remain only. The new interference coefficient $\beta_{k,l}^{q,p}$ is an altered form of $\alpha_{k,l}^{q,p}$ that depends on the difference of the CFOs for the two users, and becomes

$$\beta_{k,l}^{q,p} = \frac{\sin \pi(l - k + \varepsilon_{qp})}{N \sin\{\frac{\pi}{N}(l - k + \varepsilon_{qp})\}} \exp\left\{j\frac{\pi \varepsilon_{qp}(N-1)}{N}\right\} \exp\left\{j\frac{\pi(l-k)}{N}\right\} \tag{17.12}$$

where ε_{qp} means the CFO difference between user p and user q defined as

$$\varepsilon_{qp} = \varepsilon_p - \varepsilon_q \tag{17.13}$$

If we know all frequency offsets, ε_{qp} can be calculated easily and so does $\beta_{k,l}^{q,p}$ since it depends only on the DFT size N, subcarrier spacing $(l - k)$ and the CFO difference ε_{qp}. The residual MUI component in (17.11) is eliminated in frequency domain.

17.3.2 MUI Cancellation in Frequency Domain

To remove the residual MUI, we applied the PIC after channel equalization in frequency domain [5]. In this paper, we assumed that all the subchannel gains are known. Let the equalizer output be Z_k^q, then

$$Z_k^q = X_k^q + \sum_{\substack{p=1 \\ p \neq q}}^{Q} \sum_{l \in S_l} \beta_{k,l}^{q,p} \left\{\frac{H_l^p}{H_k^q}\right\} X_l^p + W_k'' \tag{17.14}$$

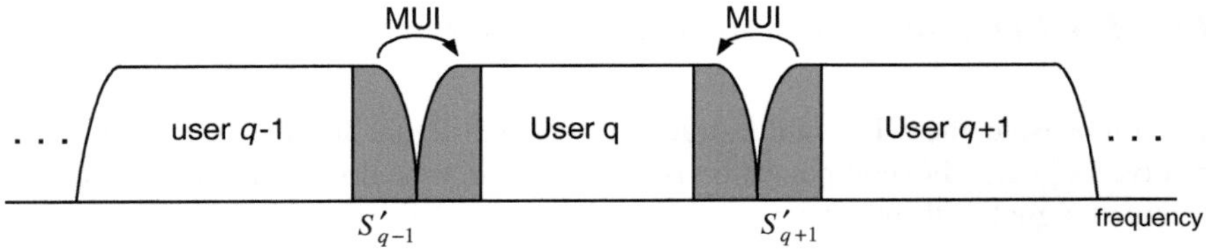

Fig. 17.1 MUI cancellation range of the proposed method with localized mapping method for OFDMA

From (17.11), in order to subtract the MUI components from the kth subcarrier signal of user q, we need to know all the correct symbols of other users, $\{X_l^p\}, p = 1, \ldots, Q, p \neq q,\ l \in S_p$. We use the decision value of $\{Z_l^p\}$ as the temporary estimate of the symbol, that is

$$\hat{X}_l^p = \mathrm{dec}\{Z_l^p\}, \quad p = 1, \ldots, Q,\ p \neq q,\ l \in S_p \tag{17.15}$$

where dec{} denotes symbol decision function. By using the initial decision and the interference coefficient of (17.12), we calculate the MUI, then subtract it from the equalizer output Z_k^q.

Since the shape of the spectrum of one subcarrier in OFDM is sinc function, the power of the interference decays exponentially as the subcarrier distance becomes larger. Therefore, we may disregard the interference from the subcarriers which are far from the desired subcarrier. In this paper, we consider the localized allocation of subcarriers for a user, and subtract the MUI for near subcarriers to both sides of the band as shown Fig. 17.1. The resulting estimated symbol is then

$$\tilde{X}_k^q = Z_k^q - \sum_{\substack{p=q-1 \\ p \neq q}}^{q+1} \sum_{l \in S'_p} \beta_{k,l}^{q,p} \left\{ \frac{H_l^p}{H_k^q} \right\} \hat{X}_l^p \tag{17.16}$$

where $\tilde{X}_k^q$ is the MUI-cancelled symbol of kth subcarrier, and S'_p is the set of subcarriers located near to both sides of the band of user q.

17.4 Simulation Results

In this section, we present the simulation results of bit error rate (BER) performance to demonstrate effectiveness of the proposed MUI cancellation scheme. We consider AWGN channel, modulation scheme of QPSK with 4 users (Q = 4) and the DFT size N of 64. That is, the subcarrier set assigned to one user is composed of contiguous 16 subcarriers. All the CFOs are assumed to be known a priori.

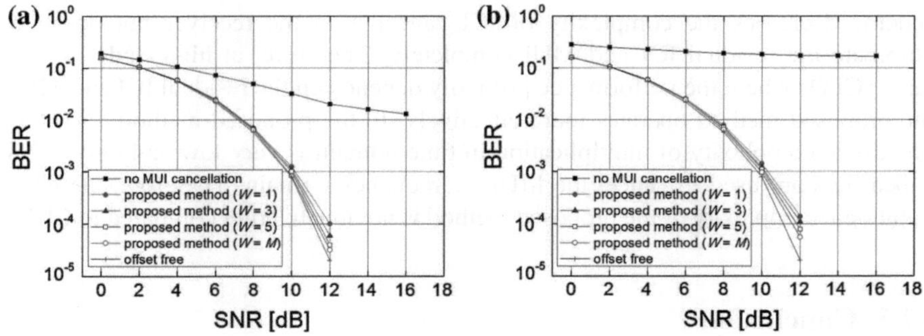

Fig. 17.2 BER performance of the proposed method with CFOs as **a** $[\varepsilon_1, \varepsilon_2, \varepsilon_3, \varepsilon_4] = [-0.1, 0.1, -0.05, 0.05]$, **b** $[\varepsilon_1, \varepsilon_2, \varepsilon_3, \varepsilon_4] = [0.1, 0.3, 0.15, -0.05]$

Figure 17.2a shows BER performance of the proposed method when the CFOs are relatively small such as $[\varepsilon_1, \varepsilon_2, \varepsilon_3, \varepsilon_4] = [-0.1, 0.1, -0.05, 0.05]$. In the figure, *W* means the number of used subcarriers to subtract the MUI. The larger *W* has better performance, and the case of $W = 5$ approaches the case, where all the subcarriers of an adjacent user is used to cancel the interference (that is $W = 16$), up to 0.06 dB and the case of offset free to 0.16 dB at BER of 10^{-4}. This means that *W* of 5 is sufficient to cancel the interference caused by CFO.

Figure 17.2b is the case where larger CFOs are used such as $[\varepsilon_1, \varepsilon_2, \varepsilon_3, \varepsilon_4] = [0.1, 0.3, 0.15, -0.05]$. The aspects of results are similar to the cases of (a) except for worse performance due to larger CFOs. In this condition, the SNR difference between the cases of $W = 5$ and $W = 16$ is 0.23 dB at BER of 10^{-4}.

The comparative results of the proposed method and CLJL scheme are shown in Fig. 17.3. In two cases, we use *W* of 5 for the proposed method. For the condition of smaller CFOs of Fig. 17.3a, the proposed method reveals 0.5 dB better performance than CLJL scheme at BER of 10^{-4}. For larger CFOs of Fig. 17.3b, the performance difference increases up to 4 dB. The CLJL scheme compensates the frequency offsets only in frequency domain after DFT by using circular convolution properties. The

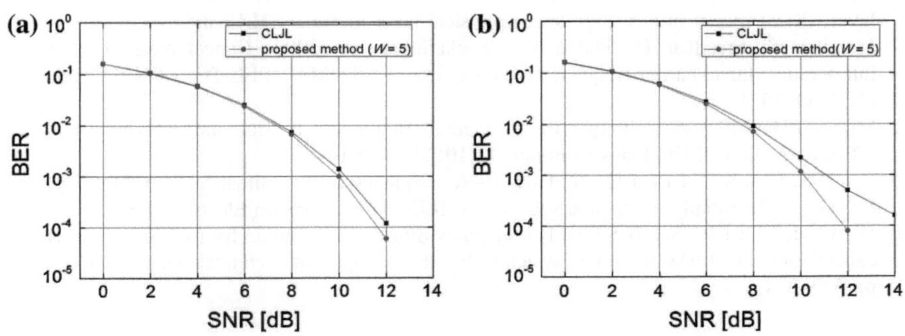

Fig. 17.3 Comparison of the BER performance of the proposed and CLJL schemes with CFOs as **a** $[\varepsilon_1, \varepsilon_2, \varepsilon_3, \varepsilon_4] = [-0.1, 0.1, -0.05, 0.05]$, **b** $[\varepsilon_1, \varepsilon_2, \varepsilon_3, \varepsilon_4] = [0.1, 0.3, 0.15, -0.05]$

scheme decreases the complexity of ICI canceller in the receiver but does not eliminate the residual ICI and MUI completely. Therefore, at high SNR or with larger CFOs where the performance primarily depends on the residual ICI and MUI, the proposed method operates more effectively. In the proposed method, the computational complexity of multiplication in time domain is very low and only a few subcarriers are used to cancel the MUI in frequency domain. Therefore, the computational complexity of the proposed method is not much larger compared to CLJL.

17.5 Conclusion

In this paper, we propose an MUI cancellation scheme for OFDMA uplink systems with multiple frequency offsets. In the scheme, we use two steps of process: CFO compensation in time domain and MUI cancellation in frequency domain. For MUI cancellation, we use a partial PIC where a few subcarriers near to both edges of the signal band are used to reduce computational complexity resulting in negligible performance degradation. Moreover, the results show that the performance of MUI cancellation is remained similarly even with large MCFO.

References

1. Bingham JAC (1990) Multicarrier modulation for data transmission for data transmission: an idea whose time has come. IEEE Trans Commun 28(5):5–14
2. Choi J, Lee C, Jung H-W, Lee Y-H (2000) Carrier frequency offset compensation for uplink of OFDM-FDMA systems. IEEE Commun Lett 4(12):414–416
3. ETSI Std. ETS 300 401 (2001) Digital audio broadcasting (DAB) to mobile, portable and fixed receiver, May 2001
4. ETSI Std. ETS 300 744 (2001) Digital video broadcasting (DVB-T); frame structure, channel coding, and modulation for digital terrestrial television, Dec 2001
5. Hou W-S, Chen B-S (2005) ICI cancellation for OFDM communication system in the time-varying multipath fading channels. IEEE Trans Wirel Commun 4(5):2100–2110
6. IEEE Std. 802.11a (2001) Part II, wireless LAN media access control (MAC) and physical layer (PHY) specifications: high-speed physical layer in the 5 GHZ band
7. Manohar S, Sreedhar D, Tikiya V, Chockalingam A (2007) Cancellation of multiuser interference due to carrier frequency offsets in uplink OFDMA. IEEE Trans Wirel Commun 6 (7):2560–2571
8. Moose PH (1994) A technique for orthogonal frequency division multiplexing frequency offset correction. IEEE Trans Commun 42(10):2908–2914
9. Seo B-S, Kim S-C, Park J (2002) Fast coarse frequency offset estimation for OFDM systems by using differentially modulated subcarriers. IEEE Trans Consum Electron 48(4):1075–1081
10. Won Y-J, Kim J-W, Seo B-S (2011) Frequency offset compensation for multiuser interference cancellation in OFDMA uplink systems. In: IEEE digest of technical papers ICCE 2011, pp 383–384

Chapter 18
Malware-Proof Embedded Systems

Robert Fitz, Wolfgang A. Halang and Lichen Zhang

Abstract Conventional measures do not sufficiently protect embedded systems against intruders and malware attacks of any kind. The main reason for this is that the system architectures are based on highly insecure and error-prone foundations. Whereas some time ago this shortcoming could still be partially coped with by swift counteraction, due to the fast data networks there are no reactive measures anymore that could compensate for the aggressors' temporal advantage. Since computers employed for purposes of safety-related automation and control or in critical infrastructures are more and more connected to networks and are, thus, endangered by malware, new architectures for their hardware and software as presented here are necessary, which completely solve the security problems by their intrinsic properties.

Keywords Embedded systems · Safety-related automation · Critical infrastructures · Malware attacks · Intrusion prevention · Hardware-based security · Security by design

18.1 Introduction

It has become fashionable to employ even for safety-related tasks in automation technology computers whose hardware and software are neither secure against intruders nor able to provide acceptable real-time performance. Thus, to avoid

R. Fitz
Faculty of Electrical Engineering, Hochschule für Angewandte Wissenschaften Hamburg, 20099 Hamburg, Germany
e-mail: fitz@etech.haw-hamburg.de

W.A. Halang (✉)
Fernuniversität in Hagen, 58084 Hagen, Germany
e-mail: wolfgang.halang@fernuni-hagen.de

L. Zhang
School of Information Engineering, Guangdong University of Technology, 100, Waihuan Xi Road, Panyu, Guangzhou, China
e-mail: lczhang@sei.ecnu.edu.cn

J.J. (Jong Hyuk) Park et al. (eds.), *Future Information Technology - II*,
Lecture Notes in Electrical Engineering 329,
DOI 10.1007/978-94-017-9558-6_18

conversions and to minimise times necessary to become acquainted with adequate industrial systems, more and more automation applications are implemented on the basis of cheap PCs as control computers and the popular Windows operating systems. As such computers are swamped with attacks for already some time now, there is a considerable risk also for industrial computing systems to be infected by malware like, for instance, Stuxnet [7] or Flame [2] and, thus, to become unsafe. This is exacerbated by the presence of almost any enterprise in the Internet, and since firewalls are unable to protect the intranets of these enterprises against external attacks.

Primarily the fast, high-capacity global communication networks and the monoculture in hardware and software technology has led to this situation, which favours the swift spreading of malware. When an electronic intruder was detected in former years, the companies dealing with counteracting them usually had sufficient time to update their products. Owing to the fact that customary software products can provide just a certain degree of protection against already known and analysed electronic malware, most computers are defenceless in the hands of new, not yet sufficiently analysed destructive programs. As there are some tens of thousands new ones of such programs every day according to studies of Bundesamt für Sicherheit in der Informationstechnik (German Federal Agency for Security in Information Technology) [1], some experts now recommend to update the installed "antivirus software" already on an hourly basis, in order to be able to provide, at least, a certain "basic level of protection".

Owing to the system homogeneity mentioned above and the increased speed of the proliferation of malware, by now it is a generally accepted fact that trying to warrant security with malware detection programs and firewalls is not an adequate solution anymore. Hence, the security problem must be solved in a fundamentally different way by appropriate architectures of hardware and software. To this end, constructive security measures are presented in this chapter, which render the malware problem manageable and, thus, contribute to the ultimate solution of this kind of security problems. The feasibility of building systems which can match the contemporary potential of threat will be shown constructively. Moreover, it turns out that such systems can even be maintained more easily as well as can provide higher performance and greater robustness as the automation systems presently prevailing.

An analysis of the various intruders, particularly in form of programs and executable Internet content with malicious intentions, reveals that they are based on some common principles of operation. If these operation principles are thwarted by appropriate measures, malware is prevented from spreading and from launching its destructive effects. The security measures presented in the sequel disable the operation principles of all known malevolent programs in effective ways. In devising them, great importance was attached to the presented solutions being simple and easy to duplicate, in order to be understood and applied without any problems by the users of computers, as unnecessary complexity is the enemy of any effort towards enhancing security.

Recently discussed approaches based on cryptography can be ruled out because of their lacking verifiability and unnecessary complexity, in particular for use in safety-related automation and embedded systems. Their benefit for the users in improving the security of conventional systems is very doubtful in consideration of the fact that there is no practically applicable cryptographic method known which could not be deciphered by attackers—let alone the costs incurred and the performance absorbed by encoding and decoding data. Moreover, in the past experience has shown that cryptographic solutions provoke playfulness, and even persons without malicious intentions feel urged to decipher systems protected this way: a kind of competition or popular sport has emerged.

18.2 Memory Segmentation

Software with malicious intentions often interferes with application programs or even with operating system routines in order to manipulate them for its destructive purpose or to deactivate software-implemented security functions. Here a memory segmentation measure as developed in [3] takes effect. It reliably prevents unauthorised accesses to the storage areas of operating system and application programs. To this end, a hardware-supervised segmentation of memory is introduced, which protects programs against modifications not permitted. The mass storage of a computing system must, accordingly, be partitioned into at least two segments. At least one of these segments has to be provided with a hardware-implemented write-protection to allow for the storage of safety-related programs and data such as operating system, utility programs and their databases, or fixed nominal values for operation and devices whose failure could lead to the destruction of devices. As shown in Table 18.1, in further segments not protected this way data are stored which, according to experience, are subject to frequent changes. At the same time, these segments can be used to test programs. This protection needs to be ensured throughout all storage levels.

Table 18.1 Hardware-supervised segmentation of memory

	Security area	Data area
Content	Operating system, utilities with their databases, fixed nominal values for operation and devices	Data
Feature	Safety-related programs and data	Not safety-related data being frequently modified
Hardware-implemented write protection	Yes	No

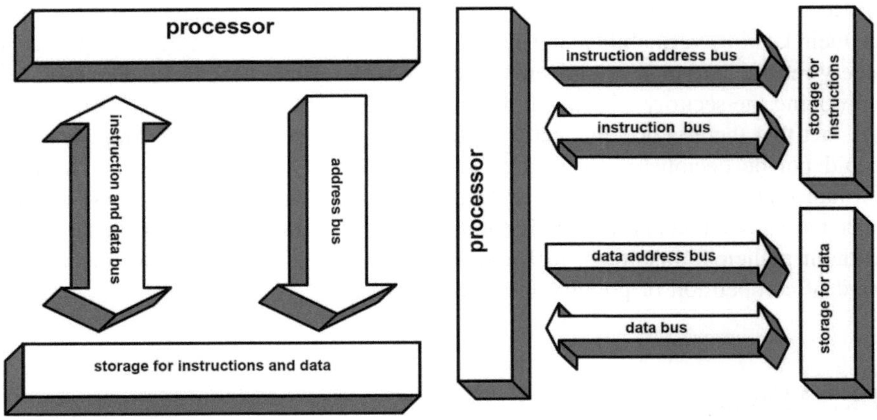

Fig. 18.1 Von Neumann (*left*) and Harvard (*right*) architectures

The now for two thirds of a century predominant Von Neumann architecture (see Fig. 18.1 left) with its minimalistic principles is totally inadequate for systems that need to be safe and secure, as it does not separate data from instructions and, thus, does not permit to protect both kinds of information in an optimum way. The actually slightly older Harvard architecture (see Fig. 18.1 right), on the other hand, provides this separation throughout and, therefore, represents an adequate construction principle. It is a pleasant side-effect that systems based on this architecture are faster than the currently prevailing ones.

18.3 Context-Sensitive Memory Allocation

In contrast to programs, data are subject to frequent modifications. Therefore, a hardware-implemented write-protection as in [3] is not feasible for handling reasons. Data can be protected against programs for spying out and modification, however, by a context-sensitive memory allocation according to [4]. Applying this measure, any unauthorised access to data is precluded. To this end, a system's mass storage, in particular the data area, is further subdivided by a partitioning into context-dependent segments. In an installation mode it is precisely specified which accesses to these segments are permitted to the programs. This is oriented at the data to be protected and not the programs, i.e. in general to each program there exist several data segments separated from one another.

In other words, this method is characterised by permitting memory references to any application program and operating system service only by means of using access functions write-protected by hardware, which release the storage areas required for the respective application case for writing and reading or just for reading accesses. Accordingly, in a hardware-protected installation mode the users must establish for any program at least one access function, if they want to use this program in the application mode. As the protection mechanism shall not hinder the users in their daily work and, in particular, shall not hamper the systems' real-time behaviour, the bounds of the memory segments assigned to the different access functions are, for instance, stored in write-protected electrically erasable programmable read-only memories (EEPROM), and loaded from there to control accesses to mass storage. Not all admissible memory areas are masked. It suffices to merely supervise the address lines and to control the write or read signals, respectively, of the mass storage media. If an access not permitted is requested, the processor is halted and a signal is generated, which allows the user to uniquely identify the incorrectly working program. For especially endangered programs a variety of access functions should be provided in order to keep the effects of infection by malware as low as possible. Electronic requests arriving from the outside, for instance, always ought to be placed first with their attachments, if any, in a separate and enclosed data area, and processed there.

This way, a spying or modifying program having infiltrated into a data segment without permission can be denied to spread to other segments leaving possible damage narrowly bounded. Based on the segmentation measures presented, a protection against unnoticed modification of data within such a segment can reliably be implemented by already established redundancy measures. Moreover, a finely structured segmentation also protects well against the negative effects of common programming errors, and provides a basis for lucid system maintenance.

18.4 Coupling of Write-Protection to Authentication

In order not to endanger the advantages of memory areas write-protected by hardware measures during the installation phases of programs, and to ensure separation on all storage levels throughout, it is necessary to accommodate service programs and their databases also in areas write-protected by hardware and separated from the program area. In doing so, it must be prevented that program and service areas are enabled for writing at the same time, and the memory management must be extended in such a way that the virtual addresses can be used to supervise the computer, since such addresses are linear and, thus, much easier to observe. This is achieved by utilising a hardware device according to [5] generating a write enable signal, which inherently prevents that more than one such signal is generated at a given instant.

For this it is necessary to ensure a unique and safe authentication of the user, which is dependent on this person's momentary function, and by means of which the access rights required for the computer's protection are selected. This implies that these systems do not designate omnipotent administrators with rights, which cannot be controlled or are extremely difficult to protect, as they always proved to be a considerable weakness in a vast number of systems under different operating systems. Therefore, almost all attackers seek to gain administrator rights in order to exercise complete control over a system. This possibility is constructively excluded in the here presented solution, since there is a kind of self-supervision of the correspondingly structured systems at any point in time. To be more precise, hardware-protected and, thus, by software not attackable components of these systems control the respective other parts, even in installation phases. Suitable for user authentication are those methods which cannot be influenced by programs and which are, for instance, based on personal property or biometrical features. To supervise the address space of a computer, safe virtual addresses dependent on authentication and start addresses of page directories are used. The memory management unit is placed between storage and processor to protect the former against direct access by the processor. The unit is equipped with a hardware-implemented protection mechanism, which transmits the required programming signals of the processor in case of correspondingly privileged authentication, only.

18.5 Disclosure of Required Resources

Destructive programs and software-based aggression from the Internet often use components of digital systems, which they would not need for their feigned nominal functions. Here the hardware-supported security measure detailed in [6] takes effect. For instance, for program modules responsible to receive electronic mail the access to communication components must be permitted, but not for those modules which display or even interpret the messages received. This example makes clear that to fulfill their nominal functions programs do not need many of the available resources at all or, at least, not all the time, and that permanent release of all resources represents an irresponsible security risk, particularly as common programming errors without malicious intentions cannot be precluded for complex software. On the other hand, users cannot be expected to disable resources on a case-by-case basis, especially not for automation systems for which this is not possible at all. Therefore, measures need to be devised which protect the users, but do not unduly trouble or restrict them.

All these problems can be solved if any program, any interpretable file and any executable Internet content first discloses which resources it requires for execution. The disclosure of a program's nominal functions enables to install boundary values for systems and to supervise their operation. By this supervision and, at any instant, by locking all resources not needed at that time by means of hardware as shown in Fig. 18.2, it can be safely warranted that the desired nominal functionality is observed. For, the users set the limits for resource accesses in installation modes, during which application software may not access processors, memory nor communication equipment. Only after that it is possible to execute application programs under permanent hardware-supported supervision based on the constraints defined before. Hereby, not only the resource accesses are supervised, but also the execution times. Thus, the real-time capability is guaranteed. As a positive side-effect, this approach also prevents, up to a certain degree, damage caused by common

Fig. 18.2 Hardware-controlled resources

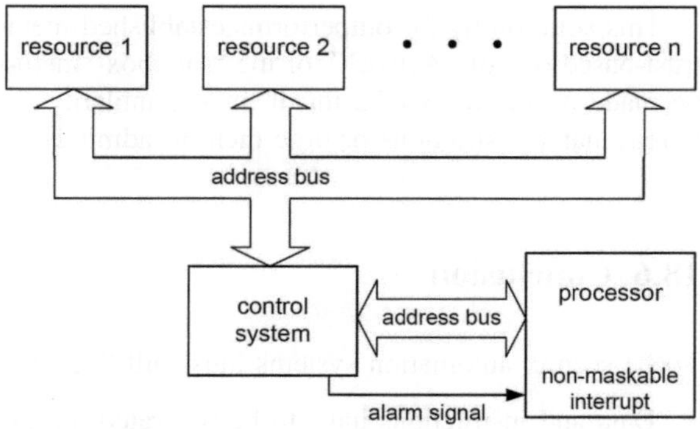

programming errors without malevolent intentions. Upon deviation from its required nominal function the corresponding program is aborted. All resources seized before are reset and released again. This has the advantage that another program can immediately be put in execution after an illicit action, i.e. the system remains available.

The here presented methods solve the problem of executable Internet contents as well, which is currently of extreme urgency. For, executable Internet contents can be considered as programs for potential spying out and modification, whose program code resides on remote computers. To cope with them, the following procedure is to be adhered to.

1. Before a program stored on a different, remotely located computer may become active, it must first provide information about its nominal functionality and the resources required for this.
2. If the intended activities are considered uncritical, the execution is initiated without asking the users unnecessary questions. What is regarded as uncritical was defined before by the users themselves, and stored in an area write-protected by hardware. Since a program's alleged activities are securely supervised in any case, the credibility of communication partners is not so decisive for a computer's security as it is the case for the currently prevailing solutions. Confidential information as exchanged, for instance, in electronic commerce is secured and encrypted for transmission here as well.
3. If the data disclosed indicate critical functions, the further proceeding depends on whether there is already a certain trust in the source of the data, and which actions were permitted to it. In case the actions requested are within the framework already authorised, here there is no feedback to the users neither. If the range of actions of an application or a data source is to be extended, first the users disconnect the communication links to extend the conceded framework of actions, and resume the connection to the communication partners not before the supervisor data and all resources not required have been hardware-protected against unauthorised access.

This solution by far outperforms established methods such as, for instance, the trust-based one of "ActiveX" or the "sandbox" method of "Java", as decisions can be made on the basis of a much finer granularity, without imposing on the users unreasonable restrictions or urge them to admit everything.

18.6 Conclusion

To be secure, automation systems must fulfill the following requirements.

- Data and instructions have to be separated throughout.
- Authentications may not be influenceable by software.
- Protection systems may not be attackable themselves. This means that their implementation must be proven correct and safely protected against modifications not permitted.
- The protection of systems may not be put out of effect during the installation phases of application programs or of operating system components as well.
- All storage levels (main memory, mass storage etc.) have to be protected throughout against unauthorised accesses by means of authentication-dependent virtual address spaces.
- Constraints and nominal functionalities of programs are defined in installation phases, and permanently supervised in the course of operation. Their observance is guaranteed even under real-time conditions.
- To protect data against effects of program errors or malicious interpretable files and to enable context-sensitive memory allocation, a means for the instantiation of programs must be provided, which employs access functions.

Utilising the presented measures, embedded systems and safety-related systems in critical infrastructures are effectively protected against inadmissible accesses. This holds in particular for still unknown attack patterns or malware, too, because there is no more need for databases of malicious code or attack prototypes, which become obsolete within hours anyway due to the swift spreading of current malware via the Internet. It has been shown that it is possible to build systems which are immune to intruders and espionage. In addition, it was shown that separation and structuring considerably facilitates the maintainability of computer control systems, too, and even increases their performance. Furthermore, it became clear that systems protected by the above mentioned measures exhibit, on the basis of disclosing their nominal functions, of the permanent supervision against set bounds, of the context-sensitive allocation of data and of the impossibility to attack operating systems and application programs, a degree of robustness which allows them to maintain their functionality despite some failing application programs—a property being of fundamental importance for embedded systems and highly safety-critical applications. The measures presented here guarantee, with reference to [8], the observance of the protection objectives *privacy*, i.e. unauthorised gain of information (spying out data) is made impossible, *integrity*, i.e. unauthorised modification of information is precluded, *availability*, i.e. unauthorised influence on the functionality is precluded, and *attributability*, i.e. the responsible persons can be identified with certainty at any point in time.

References

1. Der Schädlings-Flut Herr werden (2012). Bundeswehr aktuell 48(4):5
2. Flame: http://www.en.wikipedia.org/wiki/Flame_(malware). Looked up 26 Nov 2013
3. Halang WA, Fitz R (2000) Speichersegmentierung in Datenverarbeitungsanlagen zum Schutz vor unbefugtem Eindringen. German patent registration DE 100 31 212 A1
4. Halang WA, Fitz R (2000) Kontextsensitive Speicherzuordnung in Datenverarbeitungsanlagen zum Schutz vor unbefugtem Ausspähen und Manipulieren von Daten. German patent registration DE 100 31 209 A1
5. Halang WA, Fitz R (2000) Gerätetechnische Schreibschutzkopplung zum Schutz digitaler Datenverarbeitungsanlagen vor Eindringlingen während der Installationsphase von Programmen. German patent 10051941, since 20 Oct 2000
6. Halang WA, Fitz R (2000) Offenbarendes Verfahren zur Überwachung ausführbarer oder interpretierbarer Daten in digitalen Datenverarbeitungsanlagen mittels gerätetechnischer Einrichtungen. German patent registration DE 100 55 118 A1
7. Langner R (2011) Stuxnet: dissecting a cyberwarfare weapon. IEEE Security & Privacy 9 (3):49–51
8. Rannenberg K, Pfitzmann A, Müller G (1997) Sicherheit, insbesondere mehrseitige IT-Sicherheit. In: Mehrseitige Sicherheit in der Kommunikationstechnik. Addison-Wesley, Bonn, pp 21–29

Chapter 19
RFID Based Modular Clerk Management System in Cloud

Ming-Shen Jian, Jun-Hong Shen, Ming-Sian You, Yu-Siang Huang and Yu-Syuan Liou

Abstract In this paper, the RFID based modular clerk management system in cloud is proposed to provide the modular clerk management to different companies individually with less implementation cost. The management procedures for clerk are modular and can be adopted by the system manager individually. All the clerks are given an individual RFID tag for the induction. By managing the priority recorded in the database in cloud, the gate access management, the attendance state of clerks and even the salary for clerks can be automatically arranged. Based on cloud, different companies can access the management system and database in virtual machines. In addition, the modular management system can be used according to the on demand requirements of companies. The modular system only needs to be established once and repeatedly used for individual companies.

Keywords Clerk management · Cloud · Modular system · RFID

19.1 Introduction

Radio Frequency Identification (RFID) and Near Field Communication (NFC) technology today are the popular wireless induction systems [2, 6, 7, 9–13]. An RFID or NFC tag is given a unique ID (UID). When an independent tag approaches the antenna of RFID or NFC system, the induction between the tag and the antenna

---

This research was supported in part by Asia University under grant no. 102-asia-51.

M.-S. Jian · M.-S. You · Y.-S. Huang · Y.-S. Liou
Department of Computer Science and Information Engineering,
National Formosa University, Yunlin County, Taiwan
e-mail: jianms@nfu.edu.tw

J.-H. Shen (✉)
Department of Information Communication, Asia University, Taichung, Taiwan
e-mail: shenjh@asia.edu.tw

J.J. (Jong Hyuk) Park et al. (eds.), *Future Information Technology - II*,
Lecture Notes in Electrical Engineering 329,
DOI 10.1007/978-94-017-9558-6_19

happens. The information and content recorded in the tag is transmitted to the system. Following up the data translation, the tag recognition can be completed and related applications will be provided.

Many local or small area wireless applications for monitoring and management were proposed. For example, the RFID systems were already proposed to be used in hospital or health care [5, 14, 15]. A designed tag is given to each patient. Hence, all the patients' current location and conditions are monitored by the hospital. In other words, patients are under care even an emergency state happens.

In addition, some entrance guard systems are also based on the RFID/NFC system. The RFID/NFC ticket or card [2, 11–13] is used to identify that a user is legal or not. Some other applications focus on embedding the RFID/NFC technology into a small device such as handheld host [4]. The handheld device users can be plugged in the SD or CF interface of the RFID reader card. Hence, the users can use the RFID/NFC system everywhere.

However, the computing ability or storage spaces of the mobile devices or handheld devices are limited. In addition, individual system users may require the on demand personalized services or information data. By using the virtualization technology in cloud, virtual implementation of hardware and operating systems can be provided to users for creating or deploying their own applications with huge computing resources. Amazon's EC2, IBM's Smart Business cloud offerings, Microsoft's Azure, and Google's AppEngine are based on cloud computing [1, 3]. According to the structure of cloud computing, users' application or information data can run on virtual machines (VMs) or storage in cloud [8]. Since the virtual machine (VM) can be easily established based on cloud platform and formulated as the image file for repeat using, to provide the modular services and systems by using VMs can be an important issue.

Therefore, in this paper, a realistic application, RFID based modular clerk management system in cloud, including RFID and the modular clerk management system, is proposed. The main contributions of the proposed system are as follows:

1. Users or companies of the proposed system can be served individually. Each client user (company) will establish an individual virtual machine (VM).
2. All the client users can obtain individual or on demand defined modular services provided in virtual machines in cloud. All the information can be searched and obtained from third party applications or Internet.
3. Based on RFID, different states of different RFID users can be obtained and recorded into the database in cloud. The third party applications such as the salary system can obtain the corresponding information via Internet.
4. Based on RFID, different priorities of different RFID users can be set individually. In addition, the corresponding priority information can be adopted for the security management such as gate lock, location tracing, etc. The clerk management can be enhanced.
5. The proposed system can be embedded in other similar service systems. In addition, the modular structure and system in virtual machines in cloud can be repeatedly used. The cost can be reduced.

19.2 Proposed System

Figure 19.1 shows the service concept of the proposed system. Different companies can connect to the individual virtual machine established on cloud for system accessing. By using the modular framework provided by the proposed system, each company can create the customized content of the company services such as the website, the management system, etc. The RFID induction system and hardware are deployed at the local location. The structure of the virtual machine in cloud can be repeatedly used. All the corresponding information or content can be given a corresponding connection in the remote database. The local manager can use the web user interface for the information querying and searching. The third party application can also connect to the virtual machine via Internet for data obtaining. In addition, based on the RFID system, the individual company can enhance the clerk management system in each company individually.

Due to induction characteristics of RFID, the power supply to an RFID tag is needless. The power to drive the IC of RFID tag relies upon the induction made by the RFID antenna. In addition, each RFID tag is given a unique ID (UID) and can be used to communicate with the RFID system. Hence, in this paper, all the system users (clerks) will be given an RFID tag for identification. The system structure is

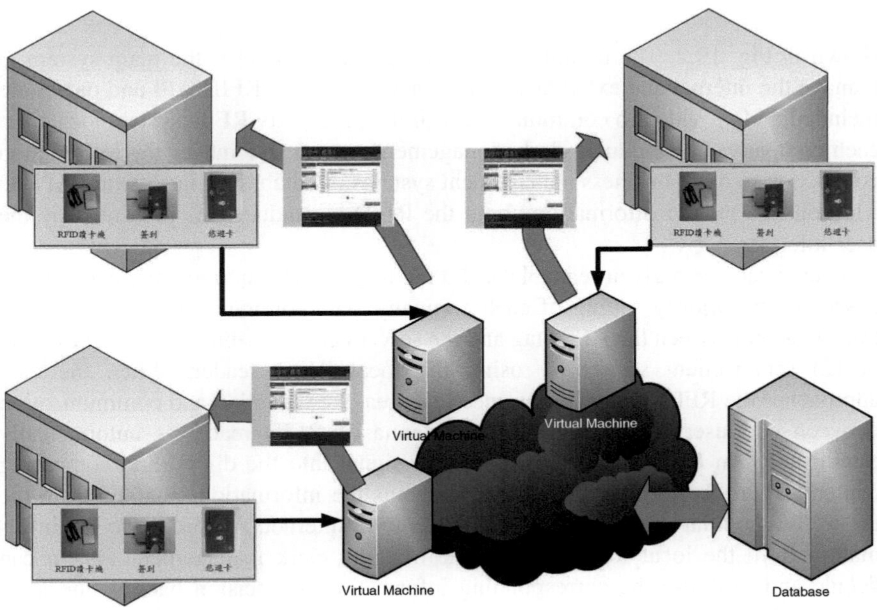

Fig. 19.1 The service concept of the proposed system

Fig. 19.2 The structure of the modular clerk management system

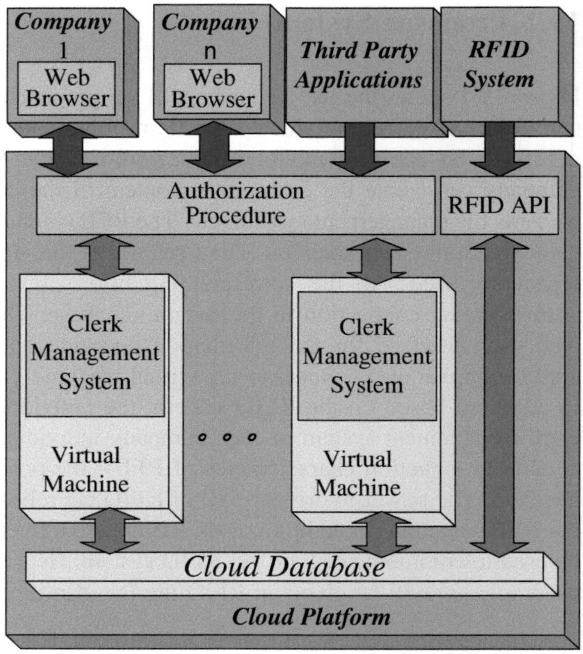

shown in Fig. 19.2. The modular clerk management system is the main system to manage the internal and external system connections. The RFID API and parser are included and provided to communicate with the third party RFID system located in each company. The modular clerk management system also makes the information connection to other business management systems or databases via software API. In addition, the related information about the RFID tag inducted is presented by the user interface.

A user can use a given readable RFID tag to gain the required personal services based on the priority setting of each company. The communication and the data transmission between the RFID tag and the server can be established via (1) Internet or (2) server-client socket by using the local RFID reader. When the user approaches the RFID system at the specified area, the induction and communication between end user RFID tag and the antenna of RFID reader is automatically established. An RFID reader will parse the signal into the digital and computing content. Then, the local RFID system transmits the information obtained from the tag to the modular clerk management system via Internet. According to the information from the local RFID system, the modular clerk management system can decide and response the corresponding information or decision back to the local RFID system or third party applications.

In addition to the local RFID system, the modular clerk management system is also implemented in the virtual machine in cloud. The modular clerk management system provides the clerk personal website user interface, the state of clerk attendance (duty of work) information and interface, and the salary interface. To modularize the framework of the clerk management system based on the database, each company can upload the corresponding information for website establishment. All the uploaded content or information will be given a corresponding file connection or data link which recorded in the remote database in cloud. Based on the connection or link information, every time the individual company creates its virtual machine in cloud, and the related information will be on demand configured for each company. In other words, even the website framework is the same as other companies, the content and information presented in the user interface will be different. It also means that the image file of the virtual machine can be created once and repeatedly used. Therefore, the cost of the application implementation can be reduced. In addition, according to the information configured by the company, the corresponding response such as the priority of the specific RFID tag can be replied to the local system or application. The related information about the specific RFID tag can be recorded in the remote database.

19.3 Implementation

To real test and verify the proposed system, the modular clerk management system is implemented in the virtual machine in cloud. The RFID readers and antennas are placed in the specific and limited local area. Considering the performance of the RFID tag, the RFID tag with size 2″ × 3.5″ for PVC and label are used. The frequency is based on the High Frequency (HF). The RFID API uses the server-client socket to communicate with the RFID reader and the virtual machine in cloud.

According to the user interface provided by the modular clerk management system, the local end user can obtain the real-time information related to the specific RFID tag via Internet. The corresponding attendance state of each RFID user can be presented via the web user interface. Figure 19.3 presents the implementation result of the web user interface for state querying. The RFID tag induction time, induction location, and the corresponding RFID user can be presented via the different color and text. Furthermore, the total time for the specific RFID tag user turning out for work can be evaluated. In other words, the salary can be counted based on the RFID induction record. Moreover, based on the RFID tag, the security in the company can be enhanced.

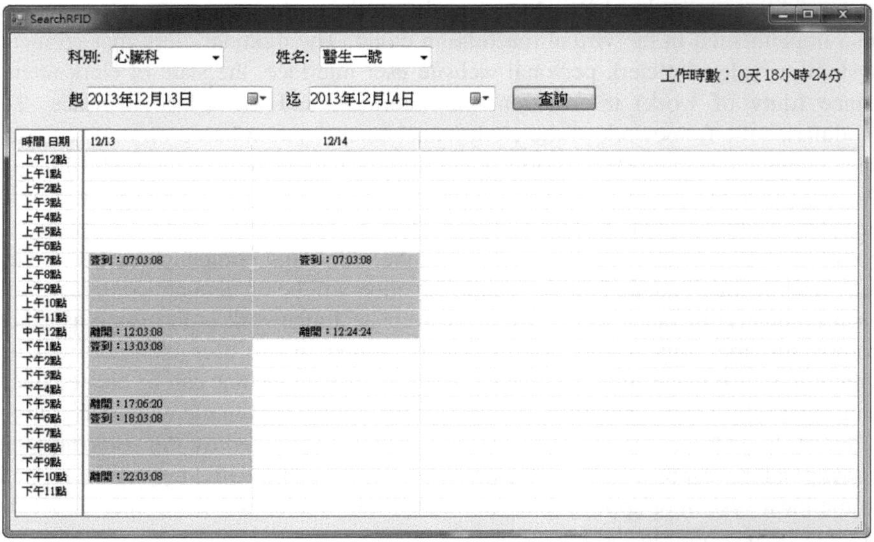

Fig. 19.3 The implementation result of the web user interface for state querying of clerk attendance (duty or work)

19.4 Conclusion

In this paper, a cloud based application, RFID based modular clerk management system in cloud, is proposed to integrate the existed management systems, RFID system, and remote database. The proposed integration system provides the RFID API module that can receive the corresponding RFID tag information from the local side RFID system. The verification shows that the proposed system is realistic and can provide the individual customized services automatically.

References

1. Amazon. Amazon elastic compute cloud (Amazon EC2). http://aws.amazon.com/ec2
2. Beijing Olympic Games Prompts RFID Development in China. http://www.rfidglobal.org/news/2007_9/200709031653253861.html
3. Christensen JH (2009) Using RESTful web-services and cloud computing to create next generation mobile applications. In: 24th ACM SIGPLAN conference companion on object oriented programming systems languages and applications, pp 627–633
4. Hoboken RFID-enables its parking permits. RFID Journal. June 2006, http://www.rfidjournal.com/article/articleview/2421/1/1/
5. Hospital uses RFID for surgical patients. RFID Journal. July 2005, http://www.rfidjournal.com/article/articleview/1714/1/1/

6. Jian MS, Lee CL, Yang KS, Huang NY (2008) Embedded automatic parking management system based on RFID and existed gate system integration. In: 12th WSEAS international conference on systems, pp 465–469
7. Jian MS, Lee CL, Yang KS, Huang NY (2008) Embedded local area public/personal service system based on RFID system integration. In: 12th WSEAS international conference on systems, pp 609–613
8. Jian MS, Jhan FJ, Chang HC, Chou TY, Shen JH (2013) Cloud based evolution algorithm with feedback control for emergency logistic. Appl Mech Mater 284–287:3370–3374
9. Keskilammi M, Sydänheimo L, Salonen P, Kivikoski M (2002) Read range analysis of passive RFID systems for manufacturing control systems. In: Proceedings of WSEAS international conferences, pp 2081–2085
10. Lu MF, Chang SY, Ni CM, Deng JS, Chung CY (2006) Low frequency passive RFID transponder with non-revivable privacy protection circuit. In: 5th WSEAS international conference on instrumentation, measurement, circuits, and systems, pp 166–169
11. Moscow metro tries RFID-enabled ticketing. RFID Journal. Feb 2007, http://www.rfidjournal.com/article/view/3049/
12. Pala Z, Inanc N (2007) Smart parking applications using RFID technology. In: 1st annual RFID Eurasia, pp 1–3
13. RFID takes a swing at ticket fraud. RFID Journal. December 2005, http://www.rfidjournal.com/article/articleview/2060/1/1/
14. RFID Hospital: Columbus children's hospital to install RFID system from mobile aspects. RFID Solution Online, March 2007, http://www.rfidsolutionsonline.com/content/news/article.asp?DocID={26FD8444-A98C-4062-911F-41F26031E4B9}&Bucket=Current+Headlines&VNETCOOKIE=NO
15. RFID trial tracks hospital equipment. http://www.computing.co.uk/computing/news/2168717/rfid-trial-tracks-hospital

Chapter 20
SMILE+: An Efficient Privacy Preserving Missed-Connection Service in IoV Networks

Shih Hui Yu, Pierre Pascal Lindenberg, Bo Chao Cheng and Huan Chen

Abstract Drivers in IoV are connected to social networks all the time and can make use of Missed-Connection Services. It provides its users with the possibility to meet strangers who were at the same place and time and allows them to contact each other at a later time. In traditional missed-connection services, the centralized server stores all information and in a scenario, where the server is untrusted, the sensitive information is in danger. We propose SMILE+, a two-phase algorithm that ensures privacy by still being protected. We also perform a privacy analysis of our approach in respect to (1) location privacy, (2) encounter privacy and (3) extra messages to show that we could make an improvement to the existing approach.

20.1 Introduction

The Internet of Vehicles (IoV) network refers to network technologies that could support highly economical and effective communication channels for the interaction between vehicles [2]. Drivers are connected to social networks all the time. Traditional social networks are based on websites, but their users' behavior has changed. Traditional web-based social networks are not convenient for mobile users anymore. However, mobile social networks (MSNs) appeares in our life. Drivers can communicate with someone who has similar interests through their IoV mobile

---

S.H. Yu (✉) · P.P. Lindenberg · B.C. Cheng
National Chung Cheng University, Minxiong, Taiwan
e-mail: saijimmy@hotmail.com

P.P. Lindenberg
e-mail: PPLindenberg@gmail.com

B.C. Cheng
e-mail: bcheng@ccu.edu.tw

H. Chen
National Chung Hsing University, Taichung, Taiwan
e-mail: huan@nchu.edu.tw

J.J. (Jong Hyuk) Park et al. (eds.), *Future Information Technology - II*,
Lecture Notes in Electrical Engineering 329,
DOI 10.1007/978-94-017-9558-6_20

devices, such as the vehicle itself or portable devices. MSNs catch the advantage from mobile networks and social networks. They support group text, location-awareness, dating service, social networking, media sharing and social gaming. Because of its features, MSNs were becoming more and more popular in recent years [3, 4, 8]. It also came into the focus of IoV applications. Missed-connection is one service of MSNs.

Websites that allow postings in area-specific groups (e.g. Craiglist.org or ppt.cc) have still many missed-connection posts on their bulletin boards. Missed-connections provide a high-tech, free version of the popular personal ads that once ran or can still be found on the back page of newspapers. Users can post a description that a person they met but failed to talk to them or exchange contact information. They hope that the person sees and replies to this post. In IoV context, a user might try to find witnesses of an accident. Missed-Connection Services (MCSs) provide the protection of anonymity rather than the risk of approaching a stranger in public while offering its users with the possibility to contact strangers who were at the same place and time at a later time. Furthermore, once connected, those strangers are able to verify to each other. However, these websites require their users to browse all the posts in hope to find one that is related to them. In case a user really finds a post that is related to him, he/she can reply or send a message directly to the poster (as shown in Figure 20.1).

While these postings offer a potential second chance for the user, his/her privacy is threatened. There are virtually no mechanisms to protect the privacy. We propose an approach to serve the users who want a second chance to meet others in IoVs. It is in relation to IoV, because people might search for persons or witnesses via missed-connection services in cases of accidents or other occurrences where the contact could not be established. Another case is when a car accident on the road and the police can use missed-connection services to find the criminals or witnesses quickly. Traditionally, people would post something on a newspaper. Nowadays,

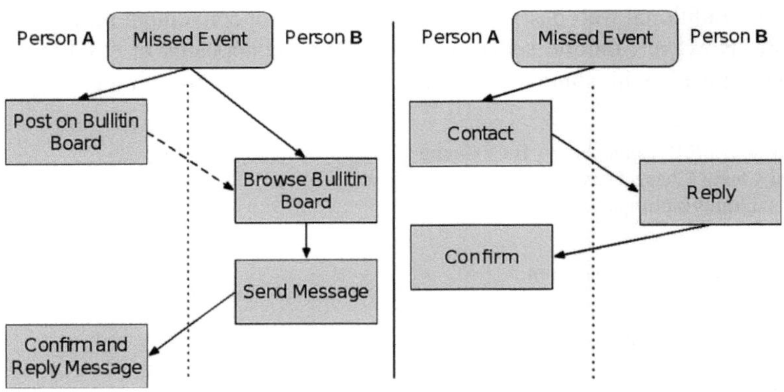

Fig. 20.1 Typical information flow of traditional missed-connection services (based on [6])

they post it on (mobile) social networks. However, this is still neither efficient nor secure. In our approach, we promote the quality of service and protect the privacy via mutual anonymity. With the high quality and high privacy protection, the police can find the criminals and witnesses quickly.

The remainder of this paper is organized as follows. In Sect. 20.2, we review the MCS scheme developed earlier. Section 20.3 contains a detailed explanation and derivation of the proposed algorithm. We provide a set of privacy analysis of our approach in Sect. 20.4. The paper concludes in Sect. 20.5.

20.2 Related Work

To understand this paper's approach, there are several things to know. The existing approach unites knowledge from k-anonymity [7] and hashing. Because it would exceed this paper's scope, only a short introduction about these will be given. Releasing tables from a relational database might contain sensitive information. Releasing these data without revealing individual identities is an issue concerning privacy. Hiding sensitive information while maintaining the best possible data integrity is a challenge [1, 5]. K-anonymity prevents re-identification by reducing the amount of the released information, so each entry appears with at least k occurrences. Hashing describes a method to take an block of data and return a fixed-size bit string. The same data will always result in the same bit string. In the following, we introduce some related works in a bit more detail. We briefly introduce Craigslist.org, what MCSs are, and we will introduce SMILE.

Craigslist is a web-based bulletin board. It provides a lot of categories, such as gigs, discussion forums, for sale, and so on. Missed-connection is one of its services. They provide web-based bulletin boards for users who want to find someone they met before but did not exchange contact information. When a user wants to find someone, he/she posts a description on the Bulletin board and expects someone to help him/her. The post is public. If there is someone who can help him/her, then he/she will reply under the post. Through the bulletin board, two strangers build a connection to each other. As seen in Fig. 20.1, the left side shows the flow of bulletin boards.

The most major problem of bulletin boards is building connections. There are too many people in the world, and it is difficult to find someone. Because the person who is the target needs to browse the bulletin board and needs to find the post, this is a hard scenario. If that person is not used to browse websites, the probability of success is approximately zero. To improve the probability, initial mobile missed-connection services appeared. It changed the behavior of missed-connections. It reduces the waiting time and increases the probability. Second, posts on Craigslist are full of sensitive information. There are location information, personal identity information, and the time in the post. All users know who has met whom where. It is too dangerous in terms of privacy. The better way is to hide the location and personal information but keep its utility.

The biggest difference between Craigslist and initial mobile missed-connection services is the waiting for an answer. While Craigslist is passively waiting, mobile missed-connection services try to find people actively. Initial mobile missed-connection services consist of a centralized server and users. When users encounter each other, they upload their personal identities, location, and time directly. With the information, the centralized server provides mobile missed-connection services to users. The centralized server is similar to bulletin boards. All of them provide a platform which lets users post their missed-connection requests. In initial mobile missed-connection services, the user stores their connections in the centralized server. When they want to have a second chance, they send a request to the centralized server. The centralized server will search the person who has the same encounter information. If there are users who have the same encounter information, they must have met the requester in the past.

However, there are several problems in initial mobile missed-connections. Users directly upload their information. In other words, they show their privacy in public. In the first case, you cannot trust the centralized server. If there is an internal abuse, then the attacker knows where the user has met whom at what time. Even though there is no internal abuse in the service provider, it still holds a risk. An attacker can steal the information from the centralized server. In another case, users directly upload their information, such as location and time that are not trustworthy. There is not authentication in the service. It is possible to fake one's identity. The initial mobile missed-connection service is the basic approach which provides the missed-connection without any protection. Attackers can steal the sensitive information easily. To solve this problem, SMILE gives a new system to protect users' sensitive information.

The authors of [6] introduce a system called SMILE, which is short for "Secure Missed Connections through Logged Encounters". SMILE is a secure, centralized missed-connections service. It assumes that (1) the service provider is untrusted and (2) the users have no pre-established social relationships with each other. SMILE uses short-range wireless communication and standard cryptographic mechanisms to mimic the behavior of users in existing missed-connections services. It provides encounter privacy protection and location privacy protection. SMILE does not upload location information and encrypts messages with encounter keys so that it provides a secure end-to-end channel. SMILE's behavior tries to be close to the behavior of people using former mentioned websites. While in public areas, the users' mobile device passively exchanges cryptographic keys with nearby peers. Updates of key hashes will be uploaded frequently to a central coordinating server. If a user wants to send a message to someone they met, he/she sends it encrypted to the server and attaches his/her encounter key. Then, the server forwards the encrypted message to all users that have uploaded the same key hash. Only users that encountered the sending user will be able to decrypt the message. This ensures that no malicious agent can determine or disclose a user's location history, encounter history, or private messages. For instance, there are two user called Alice and Bob. They met each other in subway at 1:00 p.m. They will exchange their own cryptographic keys which are private keys. In this step, they generate an encounter key (a session key) with private keys.

Upon calculating the hash value of their encounter key, the MCS application program would upload the value to the server. In other words, Alice and Bob use the hash value as a “keepsake”, for when and where they met. Since the hash value should be unique if there is no hash collision, the server is still able to derive the fact when Alice met Bob met. To solve this problem, the authors of [6] use the concept of k-anonymity. The MCS application program extracts the prefix of hash value before uploading. The prefix refers to the leading bits of whole hash value. The prefix hash values hide the unique match because of the collision. With shorter prefix, the probability of collision will be high. If the server wants to forward a message to Bob, the server will broadcast the message to all potential receivers instead. All potential users will receive the message (as shown in Fig. 20.2), but only the right user can decrypt the message. The other users might see these messages as junk messages because they cannot understand what the message means. Junk messages are just like noise, and the experience for these users is not good. It is hard to balance the quality and the privacy in SMILE. Most of time, the amount of extra messages is large. Another problem of SMILE is encounter disclosure.

SMILE supports missed-connection service via logged encounter. Even though they do not upload the time of the encounter, the centralized server still knows it, because the centralized server will log when the event occurs. As seen in Fig. 20.3, if the centralized server is not trusted then it knows the time, location and users’ ID. The centralized server knows who met whom by combining the information. To solve these problems, we propose an approach called SMILE+ that reduces the extra messages and provides a stronger protection of encounter privacy. In our approach, we keep the privacy and increase the quality at the same time.

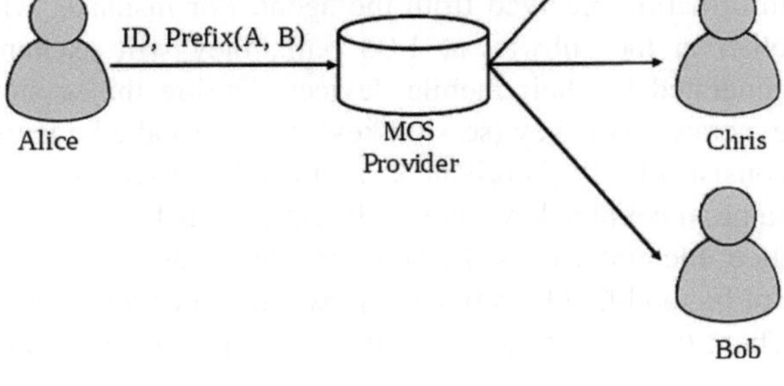

Fig. 20.2 Alice and Bob have the same prefix which is Pre(H(KA + KB)), because it is made up by the same keys

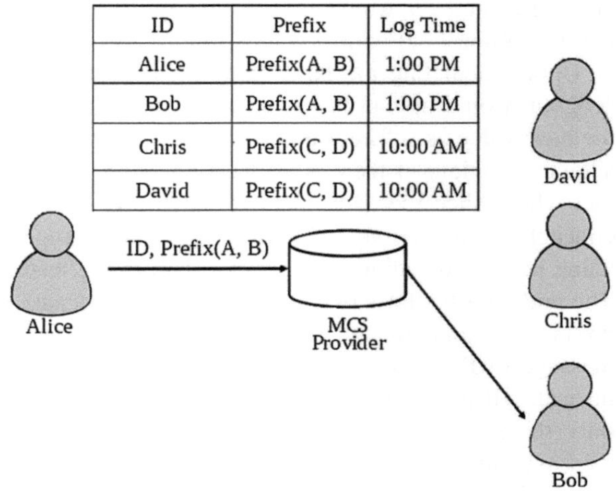

ID	Prefix	Log Time
Alice	Prefix(A, B)	1:00 PM
Bob	Prefix(A, B)	1:00 PM
Chris	Prefix(C, D)	10:00 AM
David	Prefix(C, D)	10:00 AM

Fig. 20.3 By matching the prefix with log times, the centralized server can re-identify users. The server can know who the message is directed at

20.3 Proposed Approach: SMILE+

In this section, we propose an approach using mutual anonymity to protect encounter privacy during missed-connection services. With SMILE+, missed-connection services are secure and efficient. SMILE+ consists of two phases: record connections, and rebuild connections.

20.3.1 Record Connections

When someone meets another person, but is too shy to chat, he/she can use his/her mobile device to build a connection via session key. This session key can be his/her evidence that they have met. Then, the devices compute the hash values to ensure integrity of the session keys. After that, the device will find another person who has similar interests in the mobile social network to be his/her agent. The agent will modify the sender and forward the information to the centralized server when receiving the information. In the end of this phase, the centralized server stores the information received from the agent. For instance, when Alice and Bob met each other at the subway at 1:00 p.m., they will exchange their own keys that are generated by their mobile devices. During the session, their mobile devices will generate a new key (session key) which is called encounter key. This encounter key consists of Alice's private key and Bob's private key. After Alice and Bob have the same encounter key, they will use a hash function to compute the hash values of their encounter keys. Because hashing is an irreversible function, the evidence can not be modified by anyone afterwards. Alice and Bob have the same evidence now. Then, they will find an agent to forward their messages (ID, hash(A, B), time) for

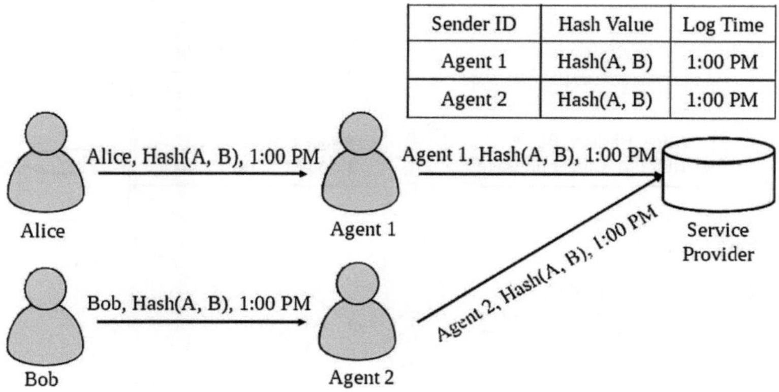

Fig. 20.4 When Alice and Bob want to record their connection

them (as seen in Fig. 20.4). The centralized server will receive the message sent by the agents instead of Alice or Bob. Thanks to the mutual anonymity, the centralized server only knows there are two users that met each other, but does not know that it is Alice and Bob. In this phase, the centralized server stores the connection between Alice and Bob. The centralized server provides the second chance to chat with each other.

20.3.2 Rebuild Connections

In this phase, we provide the second chance to someone who was too shy to chat with another person. Through this phase, he/she can have an end-to-end secure channel to the other person. When Alice wants to contact Bob again, she sends the information which includes the evidence, time and cipher messages to the agent. The evidence is the hash value of the encounter key that has been computed in previous phase. Cipher messages are encrypted by encounter key. With the cipher message, the centralized server and agents cannot recognize what message Alice wants to tell Bob. Alice and Bob can have an end-to-end secure channel between them. After the agent receives the information from Alice, the agent will record the message with three columns. First one is who send this message? Second, what is the evidence of sender? Third, when have they met each other? With these three columns, agents can query the target via the centralized server. When agents forward the information from Alice, the agents will change the sender column into the agent's name. Then forwarding to the centralized server, the centralized server will compare the time and evidences. If the evidences are the same and the time is approximately the same, the centralized server will broadcast the information to the potential users. In Fig. 20.5, agent 2 receives the information from the centralized server, and agent 2 decrypts the evidence, time and cipher messages. After comparing with the evidence table, agent

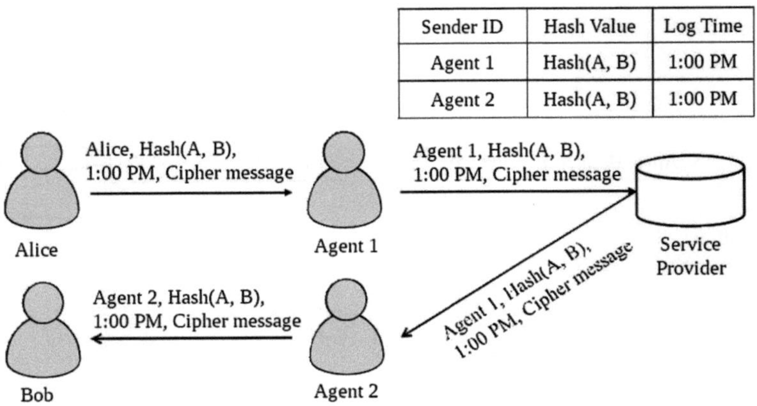

Fig. 20.5 Alice can find Bob with less extra message through this phase

2 recognizes the receiver of this information is Bob, and forwards the information to Bob with agent 2's name. In the end of this phase, Bob receives the information from Alice. He decrypts cipher messages with the encounter key. If he fails to decrypt, it means the message has been modified or the sender is another user that he did not encounter before.

According to the characteristics of hashing, different contents will result in different hash values and low repeatability. Alice can find Bob with less extra messages. For the security issue, we also extract the prefix of the hash value to be our evidence. When the prefix length equals to a maximum of the output size of the hash function (e.g., SHA-1 with 160 bits), k-anonymity will equal to one. This means that there are less extra messages in our approach, because we use the unique matches within the centralized server's table.

20.4 Privacy Analysis

In this section, we proposed our approach to archive better security and efficiency in missed-connection services. We analyze our approach in this section to show that we can protect location privacy and encounter privacy at the same time.

20.4.1 Location Privacy

Most of missed-connection services are full of location privacy leakages. For instance, the posts on Craigslist always have location information. In our approach, SMILE+ does not upload the location information. Therefore, only participants

(Alice and Bob) know the location where they encountered before. In other words, the centralized server and the agents do not know where Alice and Bob met each other. Even though the agent leaks Alice's identity, the centralized server still does not know where and whom Alice met.

20.4.2 Encounter Privacy

SMILE application program uploads the information to the centralized server without the time information. However, the centralized server's logs still can show who uploaded which information at what time, even though they use prefixes to achieve k-anonymity. As such, the centralized server is able to conclude their encounter privacy. Instead of uploading directly, SMILE+ finds an agent to work for them. With this kind of mutual anonymity, the centralized server only knows agent 1 met agent 2 but does not know the real users behind the agents. On the agents' side, the agents know the senders but do not know the receiver. The agents still do not know who has met someone somewhere. In the users' view, Alice and Bob have a secure communication channel.

20.4.3 Extra Messages

The existing approach extracts the prefix of the hash values when they protect the encounter privacy, which results in producing a large amount of the extra messages overhead due to the hash collision. When their k-anonymity is equal to four, they will produce three (k − 1) extra messages, the centralized server will find four tuples that have the same prefix in the encounter table and sends out four "rebuild connection" requests. Therefore, there are three extra messages. In our approach, SMILE+ allows the agents generating a unique match on the server side to prevent the hash collision of the encounter keys. The quality of the proposed approach is better than that of the original SMILE because SMILE+ does not make noise (messages) to other users.

20.5 Conclusion

Missed-connection service is a popular service in MSNs that gives opportunities in IoV. It provides one more chance for persons who can help the police to find people witnessing accidents. Traditional approach (Craigslist) has difficulties to find the persons that one is looking for. Initial mobile missed-connection services solve this problem but ignore the privacy concern. Another issue is that the service provider is untrusted. SMILE provides location protection and the prefixes which takes advantage of k-anonymity. However, SMILE makes noise with (k-1) messages when they achieve k-anonymity. Moreover, SMILE has the encounter privacy leakage issue potentially. From the primitive privacy analysis, SMILE+ is able to provide a secure and high quality missed-connection service environment.

References

1. Aggarwal G, Feder T, Kenthapadi K, Motwani R, Panigrahy R, Thomas D, Zhu A (2005) Anonymizing tables. In: Database theory-ICDT 2005. Springer, Berlin, pp 246–258
2. Cheng B-C, Chou C-T, Chang C-H, Lo K-H, Lin P, Hsing TR (2013) Survey on networking for internet of vehicles. J Internet Technol 14(7):1007–1019
3. Kayastha N, Niyato D, Wang P, Hossain E (2011) Applications, architectures, and protocol design issues for mobile social networks: a survey. Proc IEEE 99(12):2130–2158
4. Lane N, Walton-Flynn N, Benlamlih F (2013) White paper—mobile social networking
5. LeFevre K, DeWitt DJ, Ramakrishnan R (2005) Incognito: efficient full-domain k-anonymity. In: Proceedings of the 2005 ACM SIGMOD international conference on management of data. ACM, pp 49–60
6. Manweiler J, Scudellari R, Cox LP (2009) Smile: encounter-based trust for mobile social services. In: Proceedings of the 16th ACM conference on computer and communications security. ACM, pp 246–255
7. Sweeney L (2002) k-anonymity: a model for protecting privacy. Int J Uncertainty Fuzziness Knowl Based Syst 10(05):557–570
8. Vastardis N, Yang K (2012) Mobile social networks: architectures, social properties and key research challenges. IEEE Commun. Surveys Tutorials 15(3):1355–1371

Chapter 21
Study on the Structured Clinical Diagnosis Repository Oriented Topic Architecture

Ying Li, Yunliang Zhang, Deshan Xu, Lijun Zhu, Jie Gui and Rui Zhu

Abstract In our existing clinical diagnosis repository, as it is lack of the deeper structured knowledge, we have begun work on an integrated solution with some topic-related knowledge technologies like DITA and Topic Maps. This paper presents the new architecture of the clinical diagnosis repository, which is redesigned according to the topic-related structured knowledge techniques, as well as the technology roadmap for implementing. Applying the architecture, the repository system has been improved. This paper describes and valuates new features of the system. Finally, it presents the next-phrase tasks.

Keywords Structured knowledge · Clinical diagnosis repository · Topic · DITA · Topic maps · Topic navigation

21.1 Introduction

In China, along with more aware of knowledge economy, industry requirement for structured knowledge has become more and more urgent. For this reason, the knowledge engineering team of ISTIC is facing a great opportunity. Meanwhile, a challenge facing is to solve the problem of lack of structured domain knowledge, and provide its related services. Our goal is, to structure domain knowledge by using the topic technologies, build agile and intelligence content, to meet users' access needs of structured knowledge. On the other hand, through reusable knowledge content based-on topic, improves quality and consistency of our

Y. Li (✉) · Y. Zhang · D. Xu · L. Zhu · J. Gui
Institute of Scientific and Technical Information of China, Beijing, China
e-mail: liying@istic.ac.cn

R. Zhu
Library of Beijing Polytechnic, Beijing, China
e-mail: zhu2133@gmail.com

J.J. (Jong Hyuk) Park et al. (eds.), *Future Information Technology - II*,
Lecture Notes in Electrical Engineering 329,
DOI 10.1007/978-94-017-9558-6_21

resource content, accelerate time-to-market, reduce costs, and use the resources more productively.

As the priority domain for implementation, we choose our clinical diagnosis repository, which is getting a lot of public attention, because clinical diagnosis is related to human welfare and the quality of human life. In the repository, the data of diseases, symptoms, inspections, drugs, guides and case report etc. is being collected. Through the integration design, the optimization search, and linking the knowledge, the repository systems assists doctors for finding and obtaining relevant knowledge, and supports them in making-decision for their clinical diagnosis. In current version, it has no function that it provides users access knowledge by topics, contents in repository are not well re-used, and construction efficiency is lower. Therefore, we set up the project of "Structured Clinical Diagnosis Repository Based-on Topic" to upgrade the existing system.

This article, from the aspects of architecture and implementation roadmap, functions, implementation and evaluation, introduces the works related the upgraded system. Finally, it presents the conclusion and the next-phrase task.

21.2 Architecture and Related Technology

21.2.1 Architecture and Roadmap

Why "Structured Clinical Diagnosis Repository Based-on Topic" is important for users? Because topic-oriented content is equal to modular, component based documentation. Instead of providing large granularity content, it needed to create discrete topics of information (content resources) by users, it needed better methodologies to design, create, classify, manage, present and find content by topic for clinical diagnosis. In our study, we adopted DITA and Topic Maps standards related to topic, and designed the architecture and established the roadmap.

As shown in Fig. 21.1, the architecture contains 4 tiers:

- Collecting resources on clinical diagnosis: objects collected include medical knowledge (such as books), knowledge of diseases, cases of illness, guides/criteria, etc.
- Topicalization: structuring the collected medical resources by each topic with the DITA tool (see Sect. 21.2.2), domain ontology tool, and semantic notation tool.
- (*In DITA terminology, "content is created as topics".)
- Building the topic repository for clinical diagnosis knowledge: to assemble/generate topics into deliverables to users, defined relationships between topics, which is metadata schema.
- User interface: it provides semantic navigation function by topic that powered by Topic Maps (see Sect. 21.2.2) to satisfy individual need, and enables to deliver the contents of clinical diagnosis knowledge in various formats.

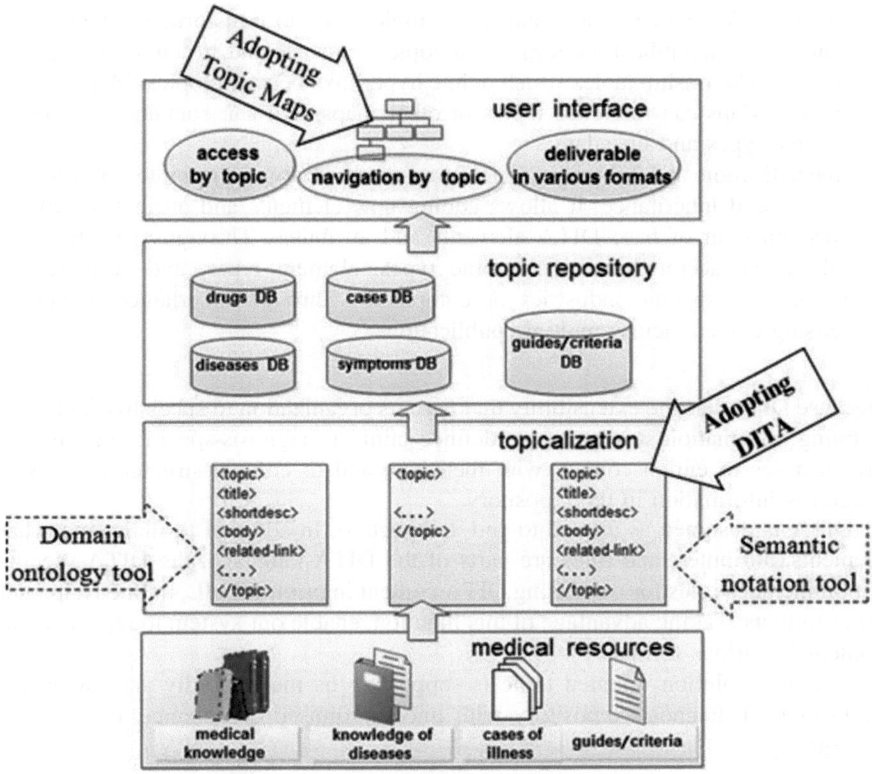

Fig. 21.1 Architecture and roadmap for implementing the structural clinical diagnosis repository based-on topic

21.2.2 Overview of DITA and Topic Maps

DITA DITA [1, 4], the Darwin Information Typing Architecture is an XML data model based-on topic for authoring and, with the DITA Open Toolkit, publishing/ presenting. It is a standard that is defined and maintained by the OASIS DITA Technical Committee. DITA has the followings characteristics:

- Topic orientation: DITA is an extensible set of structure, and its content should be structured as topics. Typically, each topic covers a specific subject with a singular intent, for example, a conceptual topic that provides an overview, or a procedural topic describing a task.
- Information Typing: DITA includes three specialized topic types: task, concept, and reference.
- Metadata: DITA includes extensive metadata elements and attributes, both at topic level and within elements.

- Maps: A DITA map is a container for topics used to transform a collection of content into a publication. It gives the topics sequence and structure. A map can include relationship tables which define hyperlinks between topics. Maps can be nested. Maps can reference topics or other maps, and can contain a variety of content types and metadata.
- Specialization and Inheritance: DITA uses the Darwin principles of specialization and inheritance. It allows adding new elements and attributes through specialization of base DITA elements and attributes. Through specialization, DITA can accommodate new topic types, element types, and attributes as needed for specific industries or companies. Through inheritance, it allows reusing content across multiple publications.

Because DITA has the extensibility that permits organization to specialize DITA by defining information structures, we defined clinical diagnosis-specific information architectures to enrich content with metadata, and to enforce structuring clinical diagnosis information in the repository.

DITA is designed as an end-to-end architecture. In addition to indicating what elements, attributes, and rules are parts of the DITA language, the DITA specification includes rules for publishing DITA content in print, HTML, online Help, and other formats. Taking advantage of this function, enable our system to represent the content in various formats.

The new solution oriented topic is supporting us more rapidly and efficiently build clinical diagnosis repository with uncompromised compliance, consistency and quality.

Topic Maps Topic Maps [3, 6], is subject-centric and semantic model, it captures and reuses semantics, relates subjects using maps, separates subject definition from the content resources, and has robust tools for ontology design, search and navigation. Topic Maps has the followings features:

- Providing a comprehensive ontology
  - Topics: defines subjects being expressed in the documentation.
  - Occurrences: content topics that relate to each subject.
  - Associations: relationships between subjects can be used to deliver and link between the content topics.
- Determining valid usage of metadata and applicability/affectivity values
- Standardizing index and glossary

DITA and Topic Maps is a natural match. DITA, it is topic-oriented content model, can develop and reuse content, combine topics using maps, separates taxonomy and relationships from content, as well as has robust tools for authoring and publishing content. DITA has interoperability with Topic Maps. It can express Topic Maps using DITA maps via specialization (new topic types, element types, and attributes), add subjects and their relationships, classify content resources.

DITA topics can be used to document the ontology, describe each subject, and publish glossary using the Open Source Toolkit. Round-trip with XTM, LTM, CTM give best of both worlds, Topic Maps tools for building, checking, navigating, searching, querying, constraining anthologies, and DITA tools for authoring, managing, publishing, translating [2].

Based on the above description, our study is taking full advantages of DITA and Topic Maps, especially adapted topicalization function from DITA, navigation function from Topic Maps as shown in Fig. 21.1.

21.3 New Features of the System

Under the topic-oriented architecture, we have implemented the development, and the main development work includes: developing a tool for building structured knowledge repository by topic, semantic notation tools, adopting our domain ontology into the structure knowledge repository, and semantic navigations system in different granularity. The system, called the structured clinical diagnosis repository, it has the followings features:

- Structured knowledge based-on topic
- Topic reuse
- Representing/Publishing the topicalized knowledge in various formats
- Metadata description basing on our domain ontology and semantic notation tools,
- Semantic Navigation based-on Topic Maps

The followings give the overall of the new features. They are structured knowledge based-on topic, topic reuse, and representing/publishing the topicalized knowledge in various formats.

21.3.1 Creating Structured Knowledge Based-on Topic

This study, according to the DITA standard, built the structured clinical diagnosis knowledge repository based-on topic. Taking AIDS as an example, the process is as follows:

- Select relevant topic.
- Decide granularity of topic, set topic content and type, topic can be divided into conceptual, task, reference topic.
- Create topic according to content and type.
- Saves, export the related file format as the needed.

Unlike building of the existing knowledge repository, the above method is based-on topic and structuring knowledge, it can re-use of knowledge.

21.3.2 Reusing Topic

Benefits of Reuse Reuse is one of the primary motivations for the clinical diagnosis knowledge repository, the benefits of reuse:

- Efficiency: Reuse improves the efficiency of building the knowledge repository.
- Effort to create content once and reuse it many times, thereby avoiding wasted or duplicate effort.
- When content that's reused must be updated; it reduces maintenance costs by updating only one topic instead of several.
- It saves on translation costs by translating reused content only once.
- Consistency and accuracy: Reuse ensures consistency and accuracy because the reused content is the same in all locations.
- Updating only one piece of content a single time ensures that the correct changes are propagated everywhere.

It can reuse content by:

- Reusing elements by using content references
- Reusing topics
- Reusing DITA maps
- Reusing content from non-DITA sources

With a content reference (short name is conref), it can refer to an element and use that element's content in place of the current element. It can conref to topics, DITA maps, and elements so that it can reuse chunks of content from a source file in a target file. It most commonly use conrefs to reuse lists, steps, tables, phrases, sentences, or paragraphs in a topic.

As shown in Fig. 21.2, it is a coding for content reuse in the clinical diagnosis knowledge repository [5].

Fig. 21.2 An example of reusing topics

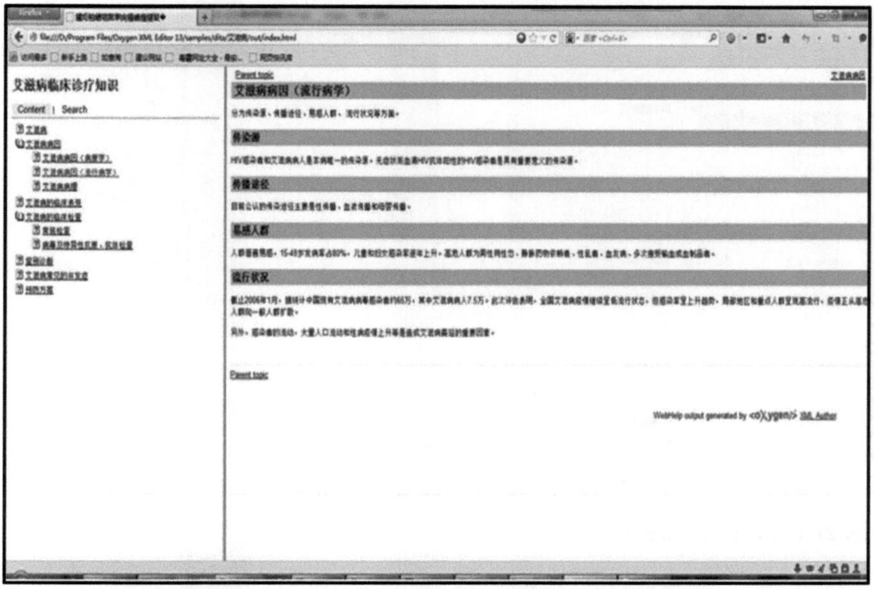

Fig. 21.3 Screenshot of resenting knowledge related AIDS in HTML

21.3.3 Representing/Publishing the Tropicalized Knowledge in Various Formats

Because this study had built the repository by adopting DITA, it can represent/publish the content by associating topics in various formats, such as html, doc, pdf, and e-put, etc. and Fig. 21.3 represents its knowledge by topics in html format for AIDS.

21.3.4 Users Interface

Disease knowledge includes overview, pathogenesis, clinical manifestation, clinical examination, differential diagnosis, common complication, clinical treatment, disease care, prevention program, guides and criterion, case-reporting. For the clinical diagnosis knowledge repository, user interface represents the knowledge by relevant topics, the interface similar to Wikipedia as shown in Fig. 21.4.

Fig. 21.4 The users' interface

21.4 Evaluation of the New Feature

Before applying the relevant topic technologies, we completed a questionnaire for the existing system. The results showed that the user satisfaction of content presentation is the lowest. As a solution, we adopted DITA and Topic Maps standards for structuring, tagging, as well as navigating contents, it have improved the user experience.

As shown in Fig. 21.5, specifically, before applying DITA, the system only has the hierarchical classification navigation, after adopting the topic technologies, we developed the association navigation among the relevant topics, such as disease, drug, and case. Meanwhile, it can export the content in a variety of formats, and reuse the required knowledge.

21.5 Conclusions and the Future Plan

Conclusions

The structured clinical diagnosis repository based-on topic reused the knowledge, improved the quality and consistency of our resource content, and reduced costs. It can be expected to accelerate time-to-market.

The future plan

In the next-phrase, as shown in Fig. 21.6, we are planning to use the tools of the ontology platform and automatic text annotation, and carry on the semi-automatic

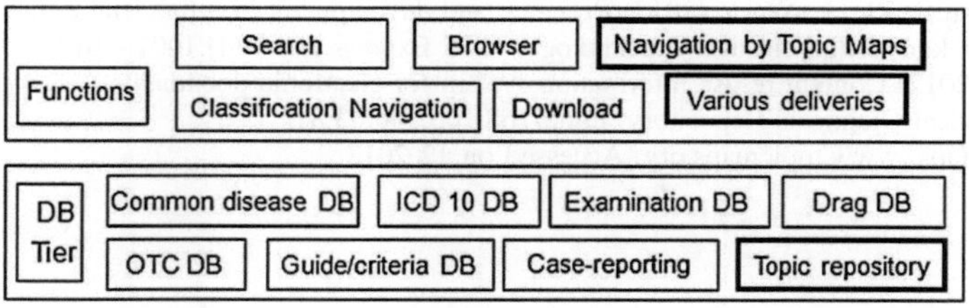

Fig. 21.5 New features shown in the *boxes* with *thicker line* are based on the topic mechanism

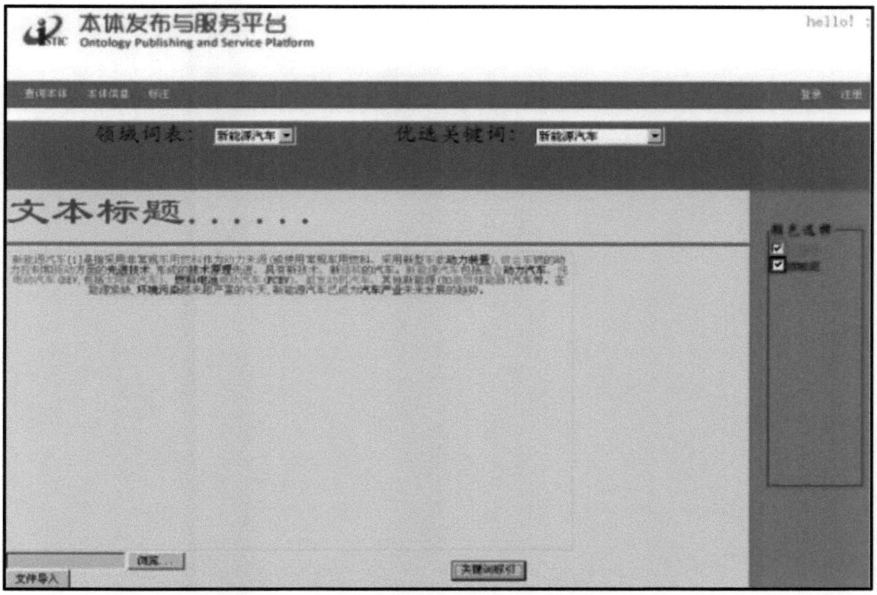

Fig. 21.6 Topic automatic indexing based-on domain ontology

tagging of topics, complete the developing the knowledge service platform based on the structured knowledge.

Acknowledgments This work is partially supported by the National Key Project of Scientific and Technical Supporting Programs under Grant No. 2011BAH10B01.

References

1. DITA OASIS Standard. http://dita.xml.org. Accessed on 12 2013
2. Gelb J (2008) DITA and topic maps bringing the pieces together. In: Topic maps conference, Oslo

3. Li Y, Zhu R, Zhu L, Wu Y (2013) Research and development of a financial e-training system based on knowledge discovery technology. ICIC Express Lett 4(4):1007–1014
4. Sanofi (2012) Content re-use information system for electronic documents
5. The Clinical Diagnosis Repository. http://168.160.200.83:81/
6. Topic Maps. www.topicmaps.org. Accessed on 12 2013

Chapter 22
The Impact of Social Capital on the Knowledge Sharing Behavior Among the Community Members of a Social Network Knowledge Platform

Chyuan-Yuh Lin, You-Shyang Chen and Yu-Wen Lo

Abstract This research has collected sample data through the social network Ask Yahoo! From the 308 effective samples, it shows that community members' attitude toward knowledge sharing certainly constitutes positive impact on the members' willingness to participate in the sharing and be loyal to the aggregative social network behavior. However, the mutual trust and common goals of the community members and their perceived social network benefits do not necessarily decide the behavioral outcomes. By taking an insight look at the behavioral traits behind people's mindset, this study aims to help understand the actual factors that influence the community members' willingness to share, so that social network operators can better run the platforms to make effective sharing mechanism and bring in tempting social network benefits.

Keywords Social capital · Perceived net benefit · Attitude toward knowledge sharing · Knowledge sharing intention

22.1 Introduction

As the Web 2.0 becomes pervasive, people are frequently using websites to share their information on a daily basis. The knowledge community members are now able to interact with one another for information exchange and discussion through

C.-Y. Lin (✉) · Y.-W. Lo
Department of Information Management, National Yunlin University of Science and Technology, 123, University Road, Section 3 Douliou, Yunlin 64002, Taiwan, ROC
e-mail: g9923807@yuntech.edu.tw

Y.-W. Lo
e-mail: g9823715@yuntech.edu.tw

Y.-S. Chen
Department of Information Management, Hwa Hsia Institute of Technology, 111, Gong Jhuan Road, Chung Ho, New Taipei 235, Taiwan, ROC
e-mail: ys_chen@cc.hwh.edu.tw

J.J. (Jong Hyuk) Park et al. (eds.), *Future Information Technology - II*,
Lecture Notes in Electrical Engineering 329,
DOI 10.1007/978-94-017-9558-6_22

the Internet. Under such circumstances, the members are interactively sharing their experiences and expertise as well as acquiring the knowledge they need. However, according to past studies, sharing knowledge with others does not match human nature [11]. Besides, anonymous interaction on the Internet signifies the unfamiliarity among the correspondents, and that will certainly hamper the willingness of knowledge sharing, although it is critical to the success of a knowledge sharing platform on the Internet. So, how to encourage the willingness of knowledge sharing is a very interesting topic worth a study. In view that Ask Yahoo! is a highly representative application which belongs to the domestic knowledge sharing platforms. The study takes it as the object for discussion on the essential factors that affect the knowledge sharing willingness of social network members.

22.2 Literature Review

22.2.1 Social Capital Theory

The social capital concept was first proposed by Bourdieu [1] based on the social activities conducted by people groups. He defined the social capital as an aggregated real or potential resources, which originated from a long and stable social network mesh that was the common capital owned by and accessible to all the members within. However, in all the prominent studies, the three social capital constructs proposed by Nahapiet and Ghoshal [10]—Structural Dimension, Relational Dimension, and Cognitive Dimension—have been the most comprehensive and complete constituent dimensions that are most often cited by social capital studies.

22.2.2 Perceived Net Benefit

The cognitive benefit and its cost is one of the most common elements discussed with knowledge sharing [12]. When an individual is making a decision, the first thing to do is evaluate the ratio between perceived effect and perceived cost, and decide whether the benefits of intangible respect and reputation, or tangible rewards will be gained [3].

22.2.3 Attitude Toward Knowledge Sharing

According to Fishbein and Ajzen [5], the attitude means an individual's positive or negative appraisal on a behavior, and the stronger the positive attitude of the individual is toward a behavior, the more likely will the individual engage in the behavior. The measurement of attitude is composed of behavioral belief and outcome evaluation of an individual. The behavioral belief indicates the significant outcomes resulted from a specific behavior by an individual, while the outcome evaluation indicates the level of importance of the significant outcomes to the individual.

22.2.4 Knowledge Sharing Intention

According to Williams [13], the intention means a tendency of desire of an individual for a future goal, and this tendency guides the individual to behave toward the fulfillment of the goal. Therefore, if the community members have good intention of knowledge sharing, they will unreservedly share their knowledge and information actively and passionately in all kinds of forms, when they are asked questions by other members. And this is the state of consistence between mentality and behavior. According to Wu and Sukoco [14], knowledge sharing can be measured by intention of participation, intention of recommendation and intention of loyalty.

22.3 Research Framework

Based on our researched topics and the literature review, we propose our research framework which is constructed by the relationships among the social capital, perceived net benefit, attitude toward knowledge sharing, and knowledge sharing intention. The framework diagram is shown in Fig. 22.1.

According to the above literature analytical induction, we assume the following:

H1 Social capital affects the attitude toward knowledge sharing.

H2 Social capital affects the perceived net benefit.

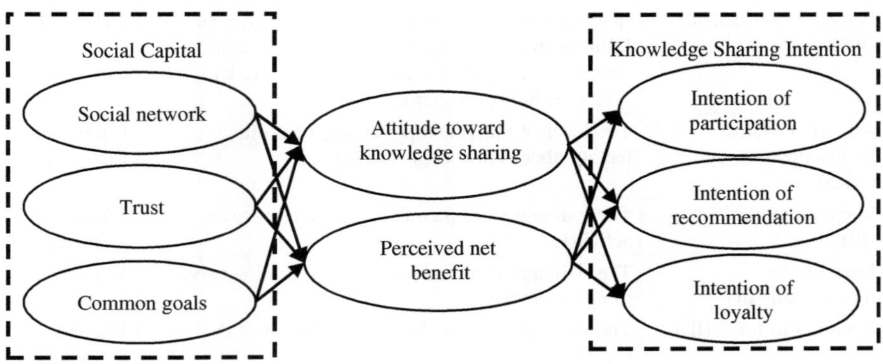

Fig. 22.1 The research framework

H3 The attitude toward knowledge sharing affects intention of individual's knowledge sharing.
H4 The perceived net benefit affects individual's intention of knowledge sharing.

22.4 Research Methodology

22.4.1 Questionnaire Design

Based on the literature review and the above assumptions, the social network factors and operational definitions are listed in the following Table 22.1, along with references to make sure the data are accountable. For the measuring variables, the Likert Scale (7-point) is adopted.

22.4.2 Data Collection

This research takes the members of the social network Ask Yahoo! as the study objects, and it uploads the questionnaire onto the My3q website (http://www.my3q.com)

Table 22.1 The social network factors and their operational definitions

Factors	Operational definitions	References
Social network (SN)	The level of connections and interactions with others	Chow and Chan [2]
Trust	The level of individual's willingness to undertake losses due to behaviors of others	
Common goals (CG)	The level of individual's common goals, missions and visions shared with others	
Perceived net benefit (PNB)	Including enhancement of work efficiency, goal fulfillment, quick response capability, external strategic relationship, service performance, and saving of human resources	Delone [4]
Attitude toward knowledge sharing (AKS)	The level of individual preference or positive feeling about knowledge sharing	Chow and Chan [2]
Intention to participate (IP)	The willingness to participate in social network activities	Wu and Sukoco [14]
Intention to recommend (IR)	The willingness to recommend the social network to non-members	Wu and Sukoco [14]
Intention to loyal (IL)	The willingness to be loyal to the brand of the social network	Wu and Sukoco [14]

and dispatches it via the PPT questionnaire board. The questionnaire is open for 2 months for people to fill out, and finally there are 308 effective copies of the questionnaire collected.

22.4.3 Data Analysis

This study uses the PLS (SmartPLS2.0) to analyze the social network's dimensions and factors. This study takes the Cronbach's α coefficient to measure the level of the variables' internal consistency. The algorithm is when the α value is greater than 0.7, it means a high internal consistency, and if it is lower than 0.35, it means a low internal consistency [7]. All the Cronbach's α values obtained in this study are greater than 0.7, which says the high internal consistency of the variables. In the verification of reliability and validity of the models used in this study, the measuring tools will be proved to have very good convergent validity and discriminant validity [9] if the following conditions are satisfied: the factor loading of each question reaches 0.5, the composite reliabilities (CR) is greater than 0.7 as recommended [8], the average variance extracted (AVE) is greater than 0.5 [6], and every factor's AVE square root is greater than the coefficients of other related factors. After the SmartPLS calculation, we've found that all the modeling factors have very good reliability and validity. The data of the model's reliability and validity are shown in Table 22.2.

After the model's reliability and validity are verified, now it comes to the examination of the true-or-false of the assumptions. The PLS model needs normalized path coefficients—the T Statistics—in order to be judged whether it reaches a statistic significance. The model's path coefficients and levels of significance are displayed in Table 22.3.

Table 22.2 The data of the model's reliability and validity

	SN	Trust	CG	PNB	AKS	IP	IR	IL
SN	0.934							
Trust	0.723	0.914						
CG	0.790	0.734	0.911					
PNB	0.699	0.748	0.701	0.935				
AKS	0.744	0.752	0.740	0.845	0.940			
IP	0.808	0.717	0.778	0.644	0.753	0.967		
IR	0.716	0.712	0.723	0.665	0.713	0.737	0.932	
IL	0.748	0.719	0.719	0.736	0.786	0.764	0.775	0.915

Note the data within the diagonals are the AVE square roots of the factors

Table 22.3 The model's path coefficients and levels of significance

	Original sample (O)	Standard error (STERR)	T statistics (|O/STERR|)
SN → AKS	−0.041	0.134	0.304
Trust → AKS	0.142	0.120	1.182
CG → AKS	0.178	0.114	1.563
SN → PNB	−0.081	0.142	0.566
Trust → PNB	0.222	0.164	1.355
CG → PNB	0.163	0.128	1.27
PNB → IP	−0.150	0.116	1.291
PNB → IR	0.070	0.173	0.405
PNB → IL	0.105	0.157	0.665
AKS → IP	0.343	0.139	2.466**
AKS → IR	0.229	0.156	1.462
AKS → IL	0.324	0.164	1.976*

*$p < 0.05$; **$p < 0.01$

22.5 Results and Discussions

Our study result shows that Ask Yahoo! has a social capital which does not have significant influence on its members' attitude toward knowledge sharing and perceived net benefit (social network, members' trust, and common goals). Although the outcome is not consistent with the results from the past researches, this study believes that this does not mean the influence is non-existent at all, because the inconsistency is attributable to the structure, goals, and interactivity of Ask Yahoo! that are different from other social networks. Here are some scenarios: If the members' backgrounds of a social network are consistent, the members will be a lot easier to get bound together tightly; if the goals of a social network are clear, the members will be easier motivated to strive for the goals; and if the interactive mechanism is good, it will be easier to enhance mutual understanding and thus creating trust among the members. But these essential elements are not easy to achieve for general social networks. It takes time and efforts to build them. As we all know, the Ask Yahoo! social network has its members from every corner of the globe, who provide infinitely versatile and knowledgeable questions and answers. How to put the people from totally different backgrounds together and build a clear vision upon a good interactive mechanism are the things to be taken cared of seriously.

Knowledge contribution is one of the main reasons behind the success of a social network, and this is especially true in the case of a knowledge sharing social network. Our study result shows that the community members' attitude toward knowledge sharing has influence on the members' intention of participation in activities and loyalty to the society. This is consistent with the results of past researches. Therefore, increasing the members' active attitude toward knowledge sharing can encourage the members to be more actively engaged in the social activities and be more loyal to the society. From the perspective of social exchange, to raise the intention of knowledge sharing, incentive motivations are indispensible. In other words, it has to let the members perceive that their contributions of knowledge will ultimately bring them benefits. Ask Yahoo! allows a member to accumulate points by participation in activities, and the more the points accumulated, the higher knowledge grade the member is awarded. This mechanism is definitely a plus to motivate its members' intention of knowledge sharing.

References

1. Bourdieu P (1986) The forms of capital. In: Richardson JG (ed) Handbook of theory and research for the sociology and education. Greenwood, New York
2. Chow WS, Chan LS (2008) Social network, social trust and shared goals in organizational knowledge sharing. Inf Manage 45(7):458–465
3. Cook KS, Emerson RM (1987) Social exchange theory. Sage Publications, Beverly Hills
4. Delone WH (2003) The DeLone and McLean model of information systems success: a ten-year update. J Manag Inf Syst 19(4):9–30
5. Fishbein M, Ajzen I (1975) Belief, attitude, intention and behavior: an introduction to theory and research. Addison-Wesley, Reading
6. Fornell C, Larcker DF (1981) Structural equation models with unobservable variables and measurement error: algebra and statistics. J Mark Res 382–388
7. Guilford J, Fruchter B (1973) Fundamental statistics in psychology and education, 5th edn. McGraw-Hill Book Co., New York, pp 503–511
8. Hair JF et al (1998) Multivariate data analysis. Prentice Hall, Upper Saddle River
9. Hair J, Black BB, Anderson R, Tatham R (2006) Multivariate data analysis. Prentice-Hall, Upper Saddle River
10. Nahapiet J, Ghoshal S (1998) Social capital, intellectual capital, and the organizational advantage. Acad Manag Rev 23(2):242–266
11. Thomas HD, Laurence P (1998) Working knowledge: how organizations manage what they know. Harvard Business School Press, Boston
12. Wang S, Noe RA (2010) Knowledge sharing: a review and directions for future research. Human Resour Manag Rev 20(2):115–131
13. Williams TH (1972) Educational aspirations: longitudinal evidence on their development in Canadian youth. Sociol Educ 45(Spring):107–133
14. Wu W-Y, Sukoco BM (2010) Why should I share? Examining consumers' motives and trust on knowledge sharing. J Comput Inf Syst 50(4):11–19

Chapter 23
Understanding University Students' Online Purchase Decisions Through Social Networking Services (SNSs)

Yan Sun, Shiqing Wang, Xue Bai and Woon Kian Chong

Abstract This study aims to investigate the effect of e-business integration with social networking services (SNSs), with a particular focus on university students' online purchase decisions. A quantitative study was adopted, using online survey approach to collect 513 valid empirical data from the university students. Data analysis was conducted by SPSS, started from descriptive aspect to regression test progressively. Several impact factors namely advertisements (ads) on web-page, ads on official homepages, ads on interest groups, and ads posted by friends associated with SNSs were identified and the usage of SNS showed significant impact on university students' online purchase decisions. The findings suggest SNSs is one of essential tool for the online purchase decision from the Chinese university students' perspective.

Keywords Social networking services · Purchase decisions · Advertisements on SNSs

23.1 Introduction

Owing to the maturity of Web 2.0 technology, the number of participants in Social Networking Services (SNSs) has a remarkable growth [9]. Claimed by Qi and Fu [11], SNS users are increasing since 2010 in China. SNSs have been explored by

Y. Sun · S. Wang · X. Bai · W.K. Chong (✉)
International Business School Suzhou, Xi'an Jiaotong-Liverpool University, Suzhou, China
e-mail: woonkian.chong@xjtlu.edu.cn

Y. Sun
e-mail: yan.sun@xjtlu.edu.cn

S. Wang
e-mail: shiqing.wang10@xjtlu.edu.cn

X. Bai
e-mail: xue.bai10@xjtlu.edu.cn

J.J. (Jong Hyuk) Park et al. (eds.), *Future Information Technology - II*, Lecture Notes in Electrical Engineering 329,
DOI 10.1007/978-94-017-9558-6_23

abundant researchers as communication tools with its interactive nature, which provides opportunities for organizations to share and exchange information, consolidate relationships with existing and potential customers, and to identify potential business problems.

The paper has been structured to outline the main purpose of the study, review the current literature on SNSs within Chinese university students. This is followed by the research methodology section. Next, the survey results and key findings are presented and discussed. The paper concludes with the limitations of the study and implications for future research.

23.2 Literature Review

SNS become a novel approach for businesses in terms of retaining with existing customers and exploring new customers [2]. In order to gain competitive advantages, many firms adopt SNSs as their "yellow paper" for expansion [12]. Additionally, it allows consumers to share and exchange experiences, which will add value when making purchase decisions [4]. Online shoppers perceived online consumers' reviews as one of the key element for purchase decision making. According to [6], university students tend to have higher acceptance rate on online shopping. For instance, university students are more willing to participate in SNSs activities because they are enthusiastic about innovation and information sharing [3]. Hence, university students are one of the most important samples for the research on both E-business and SNSs.

SNS is a powerful tool for marketing innovation [1]. SNSs provide flexible advertising approaches and efficient communication channel with interactive capability for E-Businesses [10]. Previous study has showed those consumers are tended to accept advertising on SNSs because of their social functions [12]. This study appealed that consumers' attitude to advertisement on SNSs is different owing to dissimilar type of advertisement. Furthermore, according to [8] consumer who use SNSs have more positive attitude to e-commerce. This study segmented the using of SNSs into two factors which are the using frequency and the number of contacts. It can be logically deduced that consumer who frequently use SNSs and have larger contacts number would have more positive attitude to e-commerce. Therefore, we propose the impacts that SNSs bring to purchase decision making can be categorised into three aspects: the advertisement effects on SNSs, the frequency of using SNSs and the number of contacts on SNSs.

There were enormous studies focused on SNSs and E-business, however, limited attention was paid to China—the most considerable emerging market. Besides, most prior researches ignored the significance of distinctiveness on younger generation.

23.3 Hypotheses Development

H1: Advertisements on SNSs

Specifically, this study is investigating the relationship between advertisements on SNSs and purchase decisions of university students in China. Suggested by Hoy and Milne [7], several types of online advertisement were selected and hypotheses were drafted accordingly as follows:

H1a: Ads on web-page on SNS have influence on shopping online
H1b: Ads on official home page and posted by celebrities have influence on online purchasing of electronic products
H1c: Ads on interests groups on SNS have influence on shopping online
H1d: Ads posted by celebrities have influence on online purchasing of electronic products
H1e: Ads posted by friends have influence on shopping online
H1f: Ads posted by e-friends have influence on shopping online

H2: Frequency of using SNSs

As illustrated by Brocke et al. [2], some SNS users even regard SNSs as part of their life and nearly half applied SNSs on daily basis. According to [5], the largest SNS user group in China was people aged 20–29, particularly for university students. SNS is further proved to be essential channel for extending e-marketing scale among young generation because most university students started to developing behavioural stickiness gradually. Therefore, we propose:

H2: The online frequency of SNSs has influence on online purchasing, clothes in particular.

H3: The number of contacts on SNS

Building an online circle for close friends to share all kinds of information including personal feeling and mood makes SNS more attractive to young generation worldwide. Assumedly every subscriber on SNS has an average of 150 friends, and then the company would be able to collect around 225,000 positive feedbacks if 1,500 customers agreed to help. Hence, behaviours of university students are not only impacted by their own motivations but also by others on social network. Therefore, we propose:

H3: The number of contacts on SNSs has influence on online purchasing (Fig. 23.1).

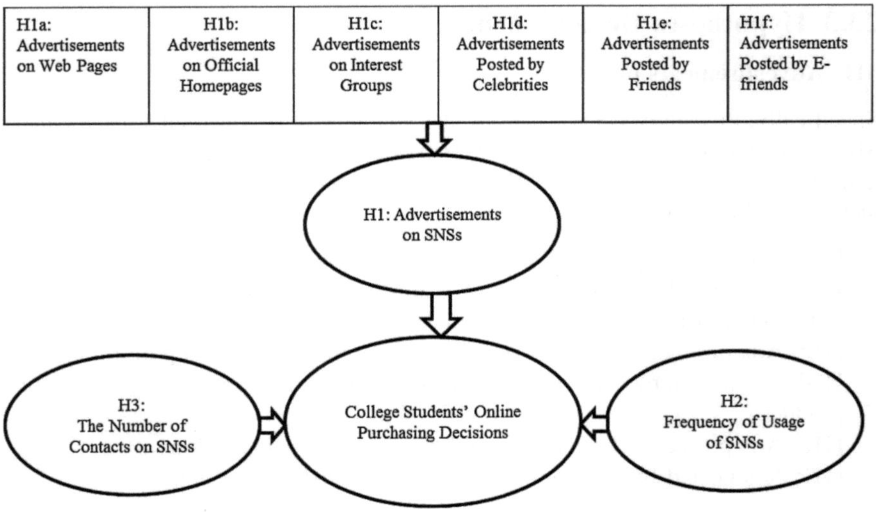

Fig. 23.1 Conceptual framework

23.4 Methodology and Survey Results

Respondents of this study are undergraduate students in colleges and universities in China. A quantitative study, using online survey was adopted. Prior to dissemination, the online questionnaire was verified by two academicians and piloted on 30 Chinese university students. Minor amendments were made to the flow and phrasing of the questions. The pilot study also confirmed the relevance and clarity of the questions to ensure the findings are consistent and relevant. 513 effective samples were obtained using the e-survey in total. Statistical Product and Service Solutions (SPSS) was used to analyse dataset and find out relationships among various variables. The full results shown in Table 23.1

23.5 Discussions

The results revealed that ads on web-page, ads on official homepages, ads on interest groups, and ads posted by friends on SNSs are positively influence the online purchase decision. University students in China seem to show higher acceptance of personalized advertisements based on their interest. Furthermore, the factor of advertising posted by friends show crucial significant level resulting from that it use the trust relationship between college students though SNSs. The research also appealed that both the frequency of usage of SNSs and the number of contacts on SNSs are positively related to university students making online purchase decision. The contacts number show the width of effect can reach through SNSs,

Table 23.1 Results of hypotheses testing

Hypotheses	Unstandardized coefficients		Standardized coefficients	T	Sig.	Results
	B	Std. error	Beta			
H1	18.064	7.733	0.102	2.336	0.02	Supported
H1a	45.69	15.986	0.124	2.858	0.004	Supported
H1b	43.438	12.816	0.147	3.389	0.001	Supported
H1c	27.295	13.71	0.087	1.991	0.047	Supported
H1d	26.337	16.443	0.07	1.602	0.11	Not supported
H1e	31.364	12.37	0.11	2.536	0.012	Supported
H1f	13.814	14.593	0.044	0.947	0.344	Not supported
H2	149.219	70.543	0.739	2.115	0.035	Supported
H3	61.837	32.397	0.083	1.909	0.057	Marginally supported

additionally; the frequency makes contribution to the depth which ads on SNSs could reach. These results lead to a conclusion that the usage of SNSs is able to affect college students in depth and width.

In practically, the results implied that SNSs can be an efficient platform for e-marketing targeting at Chinese university students owing to the high acceptance of SNSs. Choosing the appropriate advertisements on SNSs such as ads on web pages, ads on official homepages, ads on interest groups and ads posted by friends help businesses to improve marketing performance on SNSs.

23.6 Limitation and Further Research

In this research, the convenience sample posed the limitation of generalizability. Most of the participants were from large and medium-sized cities around China, where the economy and technology both developed better, as well the average consumption level and Internet penetration became higher than small cities. This means a more representative sample which includes people from more different areas with diverse background should be collected by future research.

References

1. Al-maghrabi T, Dennis C (2009) The driving factors of continuance online shopping: gender differences in behavior the case of Saudi Arabia. In: European and mediterranean conference on information systems
2. Brocke JV, Richter D, Riemer K (2009) Motives for using social network sites (SNSs)—an analysis of SNS adoption among students. In: 22nd Bled eConference eEnablement: facilitating an open, effective and representative eSociety, pp 33–49

3. Chen Y, Wu JL (2012) The influence of SNS network environment on university students and the probe of its strategies. In: IEEE computer society, pp 128–131
4. Choi HJ, Scott JE (2012) Electronic word of mouth and knowledge sharing on social network sites: a social capital perspective. J Theor Appl Electron Commer Res 8:69–82, ISSN 0718-1876 (electronic version)
5. CNNIC (2012) China Internet Network Information Center (中国互联网信息中心) (2012)
6. Doherty NF, Ellis-Chadwick FE (2003) The relationship between retailers' targeting and e-commerce strategies: an empirical analysis. Internet Res 13(3):170–182
7. Hoy MG, Milne G (2010) Gender differences in privacy-related measures for young adults Facebook users. J Interact Advertising 10(2):28–45
8. Lesma VRB, Okada H (2012) Influence of feedback from SNS members on consumer behavior in electronic commerce. In: IEEE Computer Society, pp 1189–1193
9. Li DC (2011) Online social network acceptance: a social perspective. Internet Res 21 (5):562–580
10. Nor PDKM, Nazarie WNFWM (2013) Factors influencing individuals: trust in online purchase through social networking sites. Int J Inf Sci Manage 11:1–16
11. Qi Y, Fu C (2011) The effects of flow and attachment on the e-loyalty of SNS websites
12. Richter A, Koch M (2008) Functions of social networking services. In: The 8th international conference on the design of cooperative systems

Chapter 24
User Interface Design and Evaluation for a Mobile Application with Travel News Reader

Hung-Ming Chen, Shih-Ying Chen and Jhuo-Syun Li

Abstract This study was designed around the Android mobile devices, and explored HTML5-based technology and concepts to develop an ebook reader application that can be applied to travel news reader. Current theories and methods related to the design of application interface were collected and analyzed to understand the users' satisfaction and usability regarding how applications render information. This information can be used as a reference for design of application interfaces in the future.

Keywords Android · Travel news · Ebooks · Mobile devices · User interface

24.1 Introduction

In a society with high information technology (IT), methods for how humans obtain information change because of information. In view of the fast development of IT, the provision of electronic resources has become more important. Paper resources in libraries can be divided into books, journals, theses, and newspapers. Digital resources are defined as systematic and structured processing of various data that have been digitized, then provided for people's reading and use through different devices. Therefore, digital resources include ebook, electronic journals, theses database, and e-newspapers. Ebook is an important resource in digital resources [3].

Ebooks do not necessarily need to be in the form of a "book." Ebook is the use of various devices with screens, in combination with specific software, to read digitized contents and replace previous reading of paper. Therefore, based on the definition of devices, ebooks must be used with device platforms with a screen, such as PCs, notebook computers, mobile phones, and other multimedia devices.

H.-M. Chen (✉) · S.-Y. Chen · J.-S. Li
Department of Computer Science and Information Engineering, National Taichung University of Science and Technology, Taichung, Taiwan
e-mail: hmchen@nutc.edu.tw

J.J. (Jong Hyuk) Park et al. (eds.), *Future Information Technology - II*, Lecture Notes in Electrical Engineering 329,
DOI 10.1007/978-94-017-9558-6_24

Ebook reader software in these platforms must be used to brows the content of ebooks [5].

The interface is the exchange medium between humans and operating systems. Through interfaces, users can give function commands to the system, and the system itself can use interface message transmission to tell users current progress and situation of the system. How well the interface is designed is the key factor that influences whether user operation is successful. If the interface is not designed appropriately, users may need to spend more time to learn interface operation, and easily make users feel frustrated toward the product, consequently lowering the users' intention to use [2, 4]. Including user experiences in interface design considerations (through user participation and expressing their experience) during the application development process can reduce the resistance of users toward new information systems and improve users' willingness to use [1, 7].

24.2 Methods

This study used TTNews [6] travel news reader as a foundation to explore the interface configuration of reader user interface. The progression of page configuration and interaction in the interface were integrated, and an optimal design for TTNews travel news interface was proposed.

The usability testing in this study was conducted for establishing the design configuration of the application interface for travel news readers and the application flow. Qualitative method was used to test potential problems in the default setting of the initial edition of the user interface based on user experience.

24.2.1 The Design of Usability Testing

An actual travel news reader application was produced, which was divided into three parts. In Step 1, operating flow and framework design regarding the application were planned and clearly defined. In Step 2, items of consideration for user interface design in the references were used to design the initial edition of the travel news reader application interface. In Step 3, testing and evaluation of the application were conducted using operating performance and qualitative subjective questionnaire. This is to understand user experience between user and application interaction, and find potential problems and advantages and disadvantages of the interface configuration. The results can be used as a reference to revise the interface in the future.

Table 24.1 Menu page design of the system flow structure

Operating flow process	Illustrated flow process	Depth
News content	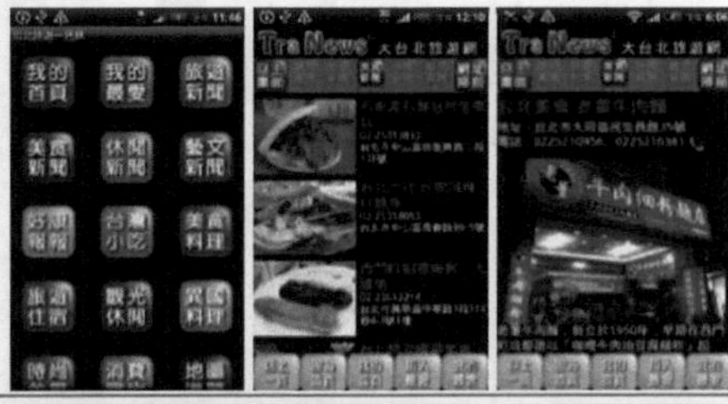	3
Change new theme	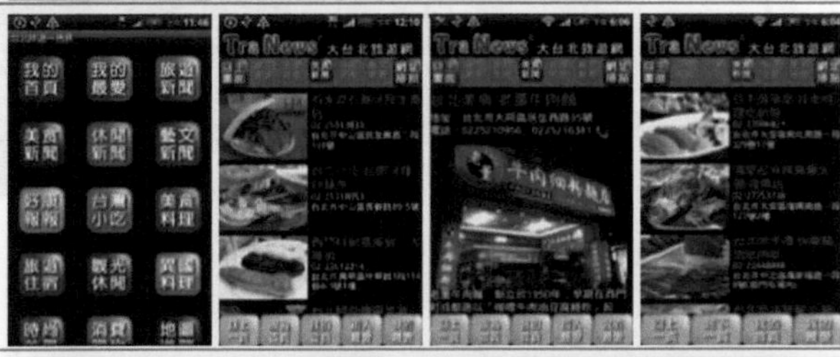	4
Toolbar functions	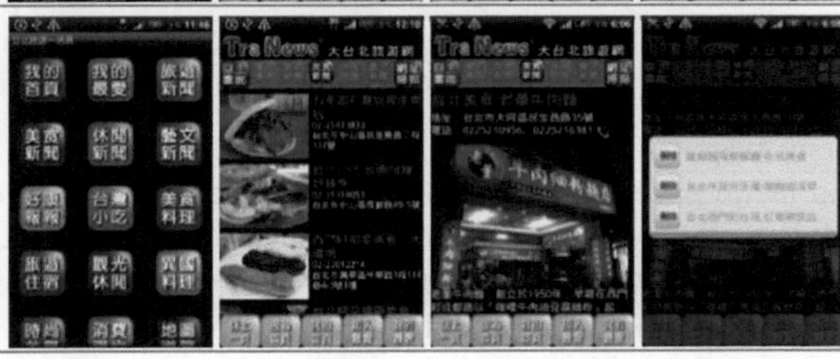	4

24.2.1.1 Planning of Application Process

The planning of the application process must clearly define the process and framework of the travel news reader application in this study, such as the width and depth of the menu. For the development of the application, this study set the total menu width of the ebook reader as 17, including 15 item buttons in the main page, and the "About" and "Leave" in the Menu. Each menu depth has a maximum of four levels, as shown in Table 24.1.

24.2.1.2 Initial Application Interface Design

The boundary of the interface in the interface configuration was the first to be set. This study used HTC Desire S with 3.7 in. WVGA touch screen and 480 × 800 (unit:

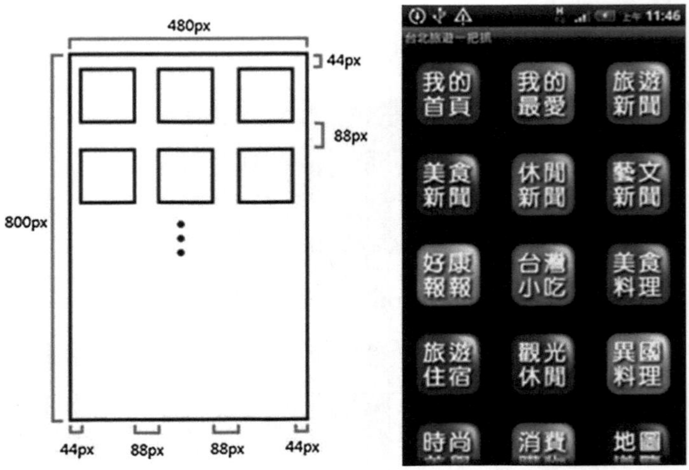

Fig. 24.1 The ratio and initial interface design of the main menu page

pixel, abbreviated as px) screen resolution to develop the travel news reader application and explore the three often used pages in the travel news reader application.

- Main menu page
  The main menu page uses 480 × 800 screen resolution as the default resolution and used dynamic method to obtain different screen resolution. Using width as a basis, the sorting method for the items in the main menu page was redistributed. The resolution for each button icon was 72 × 72, and a 44 px space was placed on the left and right side of each item. The 14 theme item buttons were sorted with an up and down scroll, as shown in Fig. 24.1.

24.2.1.3 Test of the Initial Interface

Participants in this experiment included 3 women and 13 men, with a total of 16 people. The age level is distributed between 20 and 45 years of age. The primary objective of this experiment is to evaluate potential problems and advantages and disadvantages of human-machine interaction for the initial interface. Flow structure and layout configuration of the application were also explored. Regarding the impacts on user operation, the content of the experiment primarily focuses on task operation of users by using the performance of task operation, observation during operations, and post-operation interviews.

The content of the experiment was divided into two parts, performance testing of participants' task operation, and observation of participant operation and post-operation interviews.

Table 24.2 An example of the description of tasks

Code	Function type	Task scenario	Depth
A	Browse content	Menu page switching	A-1 select travel news items from the menu of the main page. Randomly select an article from a list of 1–10 in the travel news
			A-2 select favored travel news item from the menu on the main page. In the list of 1–10, find the button for the next page, and view the 11–20 list and the 21–30 list
			A-3 select a travel news item from the menu on the main page. Find the return to main screen button in the content of the of the browse page. Select a different travel news item and continue browsing
			A-4 select a travel news item from the menu on the main page. Operate the travel news classification area in the content of the brows page and change to different type of travel news item

- Task operation
  The researchers set three tasks that allow participants to operate the interface. Users were familiarized with the basic system and system interface, and the operation and task time for each user was recorded. Researchers observed and recorded the problems encountered during the operation, as listed in Table 24.2.
- Observation and interview
  During the user's task operation, non-participation method was used to observe problems and difficulties encountered by users. When the task operation was complete, interviews were conducted to obtain the mental reflection of user during interface operation. The interviews obtained user problems, which was used as a reference for future interface revision.

24.3 Results

24.3.1 Experiment Results and Analysis

This experiment explores and analyzes an early edition of a travel news reader application interface installed on the Android smart phone. Testing and evaluation of the application used an operating performance and qualitative subjective questionnaire. This was to understand the experience users when interacting with the application, potential problems, and advantages and disadvantages of the interface and interface configuration. The results were used in the revisions of the next stage.

24.3.1.1 Experiment Participants

Participants in this experiment include 3 women and 13 men, with a total of 16 people. The ages of the participants were between 20 and 45 years old.

24.3.1.2 Experiment Design

The content of the experiment is divided into three parts.

- Basic information questionnaire
- Performance test of participants during task operation
- Observation of participant during task operation and interviews after the operation

24.3.1.3 Task Operation

Operations in this task experiment stage can be divided into three task types and 12 sub-tasks, which allow participants to conduct actual operations. Users were familiarized with this application and the application's human-machine interface. The time each user required to finish the task, from beginning to end, was recorded. The unit was in seconds.

- Task A-1: browse ebook content
- Task A-2: display inventory list 2 and 3
- Task A-3: use return to main screen button to change travel news items
- Task A-4: operate the travel news classification area and change travel news items

24.3.1.4 Task Performance Analysis

- Task A: browsing the content
  By switching the menu page, find ebooks that each individual user like and browse the content. Descriptive statistical analysis is shown in Table 24.3.

24.3.1.5 Observation and Interview

Researchers use non-participation method to observe participant during task operation to observe problems and difficulties encountered by users during tasks. After operation is complete, interviews were conducted to obtain users' mental response toward the interface during user operations and identify user problems.

Table 24.3 Task A descriptive statistics (unit: second)

Task items	Content description	Number	Average value	Standard deviation	Standard error	Min	Max
Task A-1	Browse ebook content	16	26.6	2.02	0.28	22	29
Task A-2	Display inventory list 2 and 3	16	34.8	1.85	0.19	30	41
Task A-3	Use return to main screen button and change travel news items	16	19.5	1.01	0.34	12	22
Task A-4	Operate travel news classification area and change travel news	16	22.1	1.64	0.24	19	26

After observation and interview, problems encounter by users during the operation tests for the initial interface are listed below.

- Task A-1: browse ebook content. Reading ebook content requires reading time, but the program interface does not show the current download status to users. This misleads the users in believing the delay caused by reading time was a malfunction, the application was stuck or broken, and other negative perceptions.
- Task A-2: display page two and three of the inventory list. After entering he browse content page, excessive visual elements on the screen caused the next page button to not be included in the first visual screen. The next page button was on the bottom of the list, and required the use of scrolling by touch control to find the button at the bottom of the list. Users require longer time to complete this task.
- Task A-4: operating the travel news classification area to change travel news items. When users are at the Task A-4 task, they already know how to select travel news items from the main menu page and move to the browse content page. When users proceeded with changing travel news items in the travel news classification area, finger scrolling is required by the user (which were not told to the users ahead of time). Items were displayed using gallery style sequences. When the item has been decided, selections must be made to change the content in the travel news list.

24.3.1.6 Summary of Experiment

Regarding the previously described human-machine interface problems encountered during user operation of the application, the following revision items are organized for revisions and follow-up re-testing.

- When reading the news content, users must be notified of the current progress to avoid giving users negative perceptions and second guessing the application.
- After entering the browse content page, reduce the number of visual elements on the screen and take off unnecessary information on the travel news list (telephone numbers, addresses).
- Operation of the travel news classification area to change travel news items require using fingers to slide and select reading content, and to display travel news lists.

24.4 Conclusions

This study developed a travel news reader application based on the Android smart phone. The experiment stage was used to conduct tests and evaluations of the reader application. Qualitative methods were used to obtain and record user experience and response. The results were used as the basis for revising the application interface or flow process.

References

1. Britton A, Setchi R, Marsh A (2013) Intuitive interaction with multifunctional mobile interfaces. J King Saud Univ Comp Inf Sci 25:187–196
2. Kurtcu F (2012) An analyze of high school web interface designs in terms of graphic design. J Proc Soc Behav Sci 46:5661–5665
3. Lee KH, Guttenberg N, McCrary V (2002) Standardization aspects of eBook content formats. J Comput Stand Interfaces 24:227–239
4. Nilsson EG (2009) Design patterns for user interface for mobile applications. J Adv Eng Softw 40:1318–1328
5. Rello L, Kanvinde G, Baeza-Yates RA (2012) A mobile application for displaying more accessible eBooks for people with Dyslexia. J Proc Comput Sci 14:226–233
6. The official website of TTNews. http://tw.tranews.com/. Accessed on 01 2014
7. Wong ML, Khong CW, Thwaites H (2012) Applied UX and UCD design process in interface design. J Proc Soc Behav Sci 51:703–708

Chapter 25
Multiple Moving Targets Detection Based on Hough Transform and Matched Filter

Md Saiful Islam and Uipil Chong

Abstract The most fundamental problem in radar signal is detection of a moving object or a physical phenomenon. This problem became more challenging to the signal processing and radar communities to detect multiple moving targets under severe interference. In order to effectively detect moving targets in strong noise conditions, we proposed a new method based on matched filter and Hough transform (HT) for multiple moving targets detection. At first, the noise is removed from the received signal by the matched filter with a cross correlation technique. Then, the image processing technique of the Hough transform is used. The benefits of this method include improved performance of detection, implementation flexibility, which indicates the effectiveness and robustness of the algorithm.

Keywords Multiple moving targets · Matched filter · Cross correlation · Hough transform

25.1 Introduction

The Hough transform [9] is an efficient tool for finding a line in an image. The Hough transform can be used to detect target tracks in a multi-dimensional data space that is filled with return data from the radar. Much more data can be stored and applied in this technique. This data map which may contain target trajectories can be thought of as an image. For straight line recognition the abundant literature regarding Hough transform application can be divided into two large groups depending on the parameterization used for expressing the lines. One group uses parameters c and m to express the lines [6, 15], where c is the point of intersection with the ordinate axis and m is the slope. On other hand, the most abundant group

M.S. Islam · U. Chong (✉)
School of Electrical Engineering, University of Ulsan, Daehak-Ro Mugeo-Dong, Nam-Gu, Ulsan 680-749, South Korea
e-mail: upchong@ulsan.ac.kr

J.J. (Jong Hyuk) Park et al. (eds.), *Future Information Technology - II*,
Lecture Notes in Electrical Engineering 329,
DOI 10.1007/978-94-017-9558-6_25

corresponds to those applications using another parameters ρ and θ [2, 8], where ρ is the algebraic distance from the line to the origin and θ is the angle from a vector which is normal to the line to the abscissas axis.

For finding the trajectory of a target, Carlson et al. [3–5] first applied the Hough transform using ρ and θ in search radar. After Carlson, the Hough transform has been widely used in radar such as SAR [1, 7] and ground penetrating radar [13, 14]. Since all the points in data space are mapped to parameter space one by one, computational burden of this method is very heavy that limits applications of Hough transform in radar applications. To reduce computational complexity, Liu [12] propose a modified Hough Transform algorithm by shifting the parameter space cells for radar detection. However this approach allows the limited effect of noise. In this paper, we proposed a new method for multiple moving targets detection based on matched filter and Hough transform.

The remaining of the paper is organized as follows. In Sect. 25.2, we briefly describe the literature of Hough transform regarding to our proposed method. Section 25.3 presents reduction of noise by using matched filter. Finally, simulation results are given in Sect. 25.4 to compare our proposed method to traditional methods. It proves that our method is more effective in compare to traditional Hough transform used to detect targets.

25.2 Problem Statement and Proposed Method

A line in an image can be detected efficiently by Hough transform. An arbitrary straight line can be represented by several data points in the range-time data space. The straight line can be defined by the algebraic distance p from the origin and the angle Φ of its perpendicular from the origin. The equation of line corresponding to this geometry

$$\rho = r \cos \phi + t \sin \phi \tag{25.1}$$

If we define the value of Φ from 0° to 180°, then parameters are unique for each line. This specifies that every line in the $r - t$ space corresponds to a unique point in the Hough space. Equation (25.1) is actually used for the mapping from range-time plane to Hough plane. Figure 25.1 shows the parameters for a line in r − t space. Figure 25.2 represents a view of the Hough transform of the points in Fig. 25.1. In Fig. 25.2, Hough space shows four sinusoids correspond to four data points of Fig. 25.1. Through the four data points, the point of intersection defines ρ and Φ of line in data space. The point of intersection of all of the mapped sinusoids in Hough parameter space indicates all points in range-time space does exist on a line.

Carlson et al. [3–5] applied the Hough Transform in search radar for finding the trajectory of a target. Carlson uses two thresholds named as primary threshold and secondary threshold to declare detections.

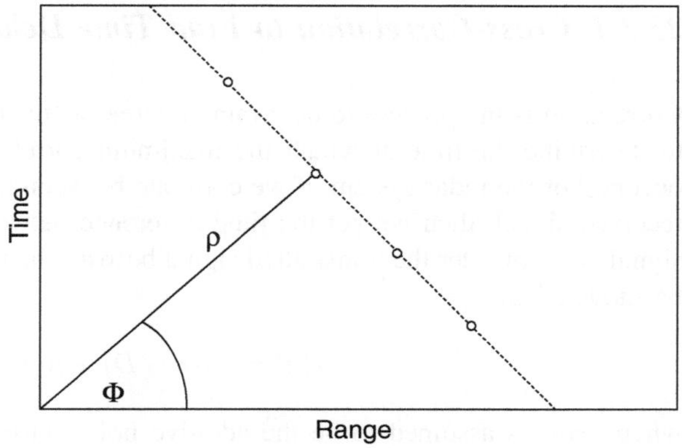

Fig. 25.1 The parameters for a line in the $r - t$ space

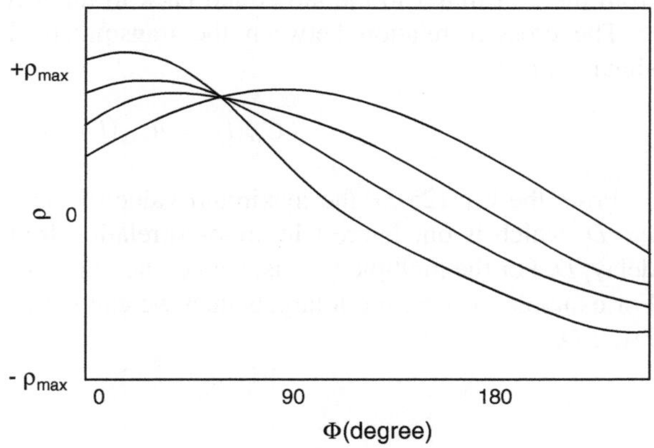

Fig. 25.2 Hough space of Fig. 25.1. Each sinusoid represents the one point of Fig. 25.1

In this paper, we employ the same method as [3] to analyze the detection performance except we use matched filter to remove noise instead of primary threshold. To remove the noise by matched filter, we use cross correlation technique to find the time delay between transmitted signal and received signal.

25.3 Noise Reduction by Matched Filter [11]

Matched filter is not a specific type of filter, but a theoretical frame work. It is an ideal filter that processes a received signal to minimize the effect of noise. Therefore, it optimizes the signal to noise ratio (SNR) of the filtered signal. So, the matched filter maximizes the SNR of the filtered signal and has an impulse response that is reverse time-shifted version of the input signal. So to obtain the maximum SNR, we need the time delay, *D*. With use of this time delay, D, we obtain the output of the cross correlation between transmitted signals and received signals.

25.3.1 Cross-Correlation to Find Time Delay [10]

Correlation is the process to determine degree of 'fit' between two waveforms and to determine the time at which the maximum correlation coefficient or "best fit" occurs. For the radar system, if we correlate between the transmitted signal and the received signal, then we get the time difference between transmitted and received signal. We consider the transmitted signal be $x(n)$, then the returned signal $r(n)$ may be modeled as:

$$r(n) = \alpha x(n - D) + w(n) \tag{25.2}$$

where $w(n)$ is assumed to be the additive noise during the transmission, α is the attenuation factor, D is the delay which is the time taken for the signal to travel from the transmitter to the target and back to the receiver.

The cross-correlation between the transmitted signal, $x(n)$ and the received signal, $r(n)$ is

$$C_{xr}(l) = \alpha C_{xx}(l - D) \tag{25.3}$$

From the Eq. (25.3), the maximum value of the cross-correlation will occur at $l = D$, which is our interest in cross-correlation from which we can get the time delay, D. For the multiple targets, we get the multiple number of D from Eq. (25.3). For example, if there are n targets then we can get n number of delay like D_1, D_2, $D_3 \ldots D_n$.

25.4 Simulation and Results Analysis

In this paper, to analyze the detection performance, we employ the same method as [3]. We assume a radar with 256 range gates and a time history composed of 128 scans with the search frame time being 0.375 s and total time history of 48 s. For this simulation, we considered three approaching targets with different radial velocity. This is shown with noise in gray-scale plot of range-time data space in Fig. 25.3.

Figure 25.4 represents the mesh plot of Fig. 25.3. Now it is almost impossible to detect the targets from Fig. 25.4 due to noise. After removing the noise using matched filter, gray-scale plot is shown in Fig. 25.5. Figure 25.6 shows the mesh plot of Fig. 25.5. From Figs. 25.5 and 25.6, it is clear that much of the noise has been eliminated. Figures 25.7 and 25.8 is the mesh plot of the Hough parameter space with noise and without noise respectively. It is almost impossible to detect the three targets from Fig. 25.7 due to severe noise. But in Fig. 25.8, we can clearly identify our three targets that is the outcome of our method.

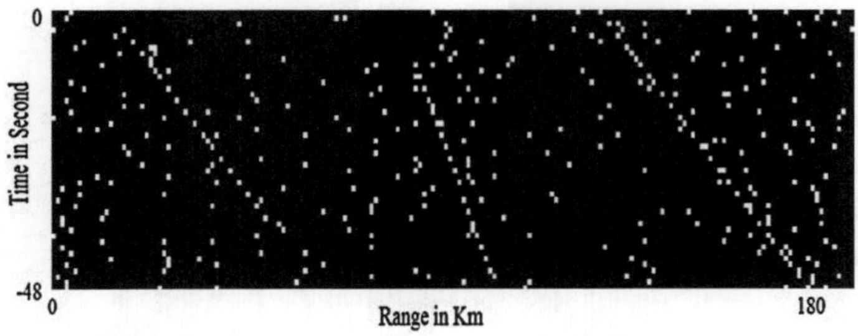

Fig. 25.3 Track of the 3 targets in the range time space. This figure represents the three track of three the targets with noise

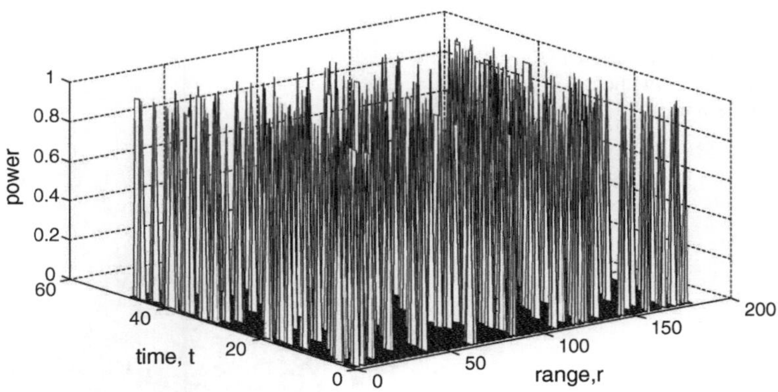

Fig. 25.4 Mesh plot of Fig. 25.3

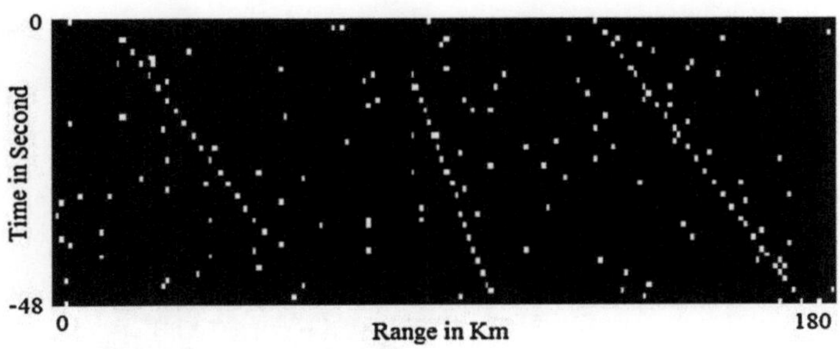

Fig. 25.5 Tracking of the 3 targets in the range time space (after noise reduction)

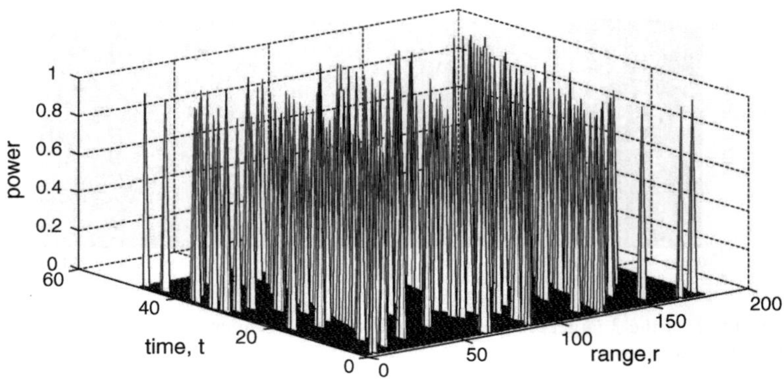

Fig. 25.6 Mesh plot of Fig. 25.5

Fig. 25.7 Hough parameter space of Fig. 25.3 (without removing the noise)

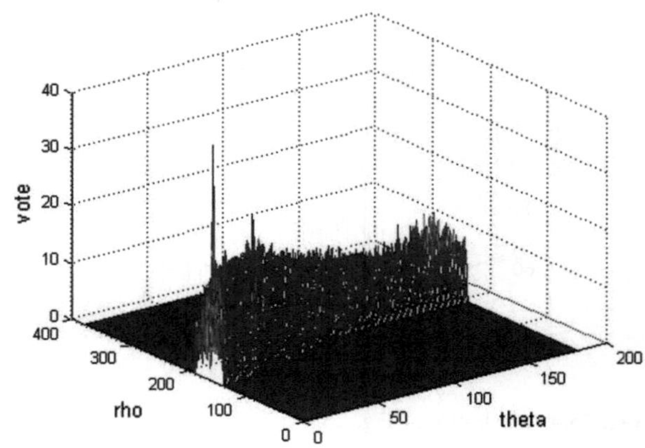

Fig. 25.8 Hough parameter space of Fig. 25.3 after removing the noise using matched filter

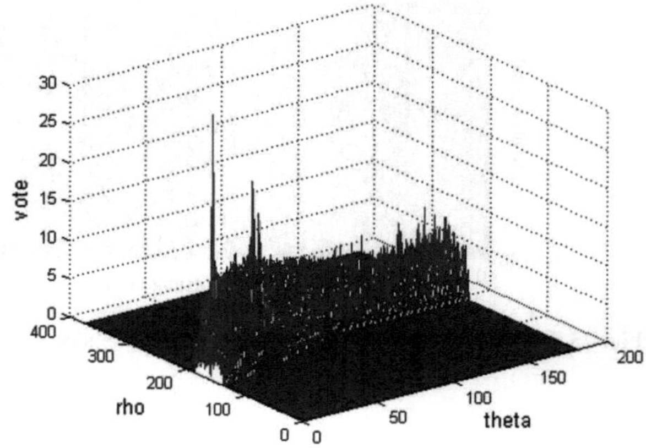

25.5 Conclusion

In this paper, we propose a new method for multiple targets detection based on Hough transform and matched filter. We use matched filter before performing the Hough transform to reduce the noise. A significant reduction in noise is achieved. The results of simulation indicate that our method is more effective in compare to traditional Hough transform used to detect targets.

Acknowledgments This work (Grants No. C0006188) was supported by business for academic-industrial cooperative establishments' funded Korea small and medium business administration in 2013.

References

1. Amberg V et al (2004) Structure extraction from high resolution SAR data on urban areas. In: Proceedings of 2004 IEEE international geoscience and remote sensing symposium, vol 3, pp 1784–1787
2. Ballard DH (1982) Generalizing the Hough transform to detect arbitrary shapes. Pattern Recogn 13(2):111–122
3. Carlson BD, Evans ED, Wilson SL (1994) Search radar detection and track with the Hough transform, Part I: system concept. IEEE Trans Aerosp Electron Syst 30(1):102–108. doi:10.1109/7.250410
4. Carlson BD, Evans ED, Wilson SL (1994) Search radar detection and track with the Hough transform, Part II: detection statistics. IEEE Trans Aerosp Electron Syst 30(1):109–115. doi:10.1109/7.250411
5. Carlson BD, Evans ED, Wilson SL (1994) Search radar detection and track with the Hough transform, Part III: detection performance with binary integration. IEEE Trans Aerosp Electron Syst 30(1):116–125. doi:10.1109/7.250412
6. Da Fontoura Costa L, Sandler MB (1989) A binary Hough transform and its efficient implementation in a systolic architecture. Pattern Recogn Lett 10:329–334. doi:10.1016/0167-8655(89)90036-6
7. Dell'Acqua F, Gamla P (2001) Detection of urban structures in SAR images by robust fuzzy clustering algorithms: the example of street tracking. IEEE Trans Geosci Remote Sens 39 (10):2287–2297
8. Duda RO, Hart PE (1972) Use of the Hough transformation to detect lines and curves in pictures. Commun ACM 15(1):11–15. doi:10.1145/361237.361242
9. Hough PVC (1962) Methods and means for recognizing complex patterns. US Patent 3,069,654
10. Islam MS, Chong U (2013) Detection of uncooperative targets using cross-correlation in oceanic environment. Research notes in information science (RNIS), vol 14, June 2013
11. Islam MS, Han H, Lee JI, Jung MG, Chong U (2013) Small target detection and noise reduction in marine radar systems. Sciverse Sciencedirect, IERI procedia
12. Liu H, He Z, Zeng J (2008) An improved radar detection algorithm based on Hough transform. Sens Imaging 9:1–7 (Springer Science+Business Media, LLC)
13. Long K et al (2006) Image processing of ground penetrating radar data for landmine detection. In: Proceedings of the society for photo-instrumentation engineers, pp 6217, 62172R. doi:10.1117/12.664038

14. Tantum SL et al (2002) Comparison of algorithms for land mine detection and discrimination using ground penetrating radar. In: Proceedings of the society for photo-instrumentation engineers, vol 4742, pp 728–735. doi:10.1117/12.479146
15. Xu L, Oja E, Kultaken P (1990) A new curve detection method: randomized Hough transform (RHT). Pattern Recogn Lett 11:331–338. doi:10.1016/0167-8655(90)90042-Z

Chapter 26
RFID-Enabled Smart Attendance Management System

Meng Zhi and Manmeet Mahinderjit Singh

Abstract The existing conventional attendance system requires students to manually provide their signature on the attendance sheet. The drawback of this conventional method is the loss of time in jotting down the attendance, leading to human error and cheating issues and the inability of capturing and storing the attendance record for other essential university activities. In this project, Radio Frequency Identification (RFID)-enabled Smart Attendance Management System (RFID-SAMS) will be designed to cater to the limitation of the conventional attendance system. The RFID-SAMS system function to enrolled students and staff data and course related information, record and monitor students' attendance and their movement during any courses attended. Its ability to uniquely identify each person based on their RFID tag type of ID card make the process of taking the attendance easier, faster and secure as compared to conventional method. The uniqueness of RFID-SAMS is the capability to detect any duplication RFID event readings and miss readings. In addition, this system has been enhanced to cater the latest ubiquitous technology such as the mobile devices, which are Window Phone and Windows Surface RT. The overall system provides an efficient, accurate and portability solution and work well for the real-life attendance environment within the university.

Keywords RFID · Attendance system · Smartphone · Tablet · Windows phone · Windows surface · Web page

M. Zhi · M. Mahinderjit Singh (✉)
School of Computer Science, Universiti Sains Malaysia, Main Campus,
George Town 11800, Pulau Pinang, Malaysia
e-mail: manmeet@cs.usm.my

M. Zhi
e-mail: zhimeng1991@gmail.com

J.J. (Jong Hyuk) Park et al. (eds.), *Future Information Technology - II*,
Lecture Notes in Electrical Engineering 329,
DOI 10.1007/978-94-017-9558-6_26

26.1 Introduction

The emergence of electronic paradigm for learning compared to traditional method and availability of almost all information on the information superhighway (Internet), nowadays have caused students to be less motivated to come to the lecture rooms than ever before. Laziness on the part of students, nonchalance to schoolwork, extra social activities that have no importance in aiding the objectives of the institution and a lot more, may prevent students from attending lectures. Sequel to these, lecturers and administrators in most developing countries have had to come up with ways to ensure a healthy participation from students, and make sure that the student-lecturer interactive relationship is kept intact. This in some cases have come in simple forms like roll calls, while in more interesting cases, can be formats like surprise quizzes, extra credit in class, etc. These strategies are however time consuming, stressful and laborious because the valuable lecture time that could otherwise been used for lectures is dedicated to student attendance taking and sometimes not accurate. In addition to all these challenges, the attendances are recorded manually by the tutor and therefore are prone to personal errors. There arises a need for a more efficient and effective method of solving this problem. A technology that can solve this problem and even do more is the RFID technology [2, 4, 5]. The existing conventional requires students to manually sign the attendance sheet every time they attend a class. As common as it seems, such system lacks of automation, where a number of problems may arise. This include the time unnecessarily consumed by the students to find and sign their name on the attendance sheet, some students may mistakenly or purposely signed another student's name and the attendance sheet may got lost [4]. For instance, in the last time our school held a career fair for students to hunt internship. This fair is compulsory for students of third and final year to attend. Then, the attendance list had to be used. In the process of students signing attendance, attendance list had to be passed one by one, if some students signed wrongly the administrator would not recognize any more, if the last student finished signing putted it aside and forgot to back to administrator, the list is easy to be lost. Therefore, the traditional attendance system has a lot of weaknesses. Besides that, the old attendance system has another drawback, which is the data of attendance list is hard to connect to other system. If the user wants to calculate the percentage of attendance or some others, he has to calculate manually or input by typing. Therefore, the RFID attendance system generated and this system can solve all problems as before.

Attendance system is one of the most important things to ensure students attend for the particular lectures. Some universities regulate class attendance as compulsory to each student who registered for a particular course. Normally in Malaysian university, students are required to attend the class not less than 80 % per semester otherwise student will be barred from taking any examinations [1]. Therefore, the attendance system is very significant in school or university [3].

The aim of this project is for improving the efficient and accuracy of students signing attendance and lecturer could view and track students' attendance. The objectives of the system are listed as below: (i) To design a webpage-enabled RFID system that functions on smartphone and tablet. (ii) To provide an effective system capable of detecting misses reading and duplication reading. (iii) Integrate Monte Carlo simulation algorithms in generating RFID EPC ID linked to students' cards information for enrollment purposes. The significant of this system is to implement a complete and novel prototype application for Windows Phone application and Windows Surface application.

The rest of the paper is organized as follows: Section 26.2 starts with related works that background of current attendance system. The methodology of this project will be introduced in Sect. 26.3. Section 26.4 is the project implementation. Finally, the discussion and conclusion present in Sects. 26.6 and 26.7 respectively.

26.2 Background and Related Work

In early edition of wireless communication systems, there are a few technology have been successfully applied in attendance system such as biometrics, barcode, smart card and etc. RFID nowadays has taken a concern in providing wireless communications with security features as well as improved the previous technology such as the communication done in wireless, fast, and easy to operate [1].

26.2.1 Biometric Time Attendance System

Biometric Time Attendance System with USB flash drive data download facility, this facility is useful for attendance data gathering when your Biometric Attendance Machine is installed on gate of office and your server or pc is placed far. LAN cable connectivity is not possible. UBS pen drive data download option will be very convenient for such case. You can collect employee attendance records from Attendance Device whenever required to generate Attendance reports [1].

26.2.2 Barcode

Barcode technology is a method of identification, which is used to retrieve in a shape of symbol generally in bar, vertical, space, square, and dots, which have different width with each one. A reader or scanners are required to identify the data that represent by barcode by using light beam and scan directly to barcode. During scanning process a scanner measured intensity of reflected light at black and white region. A black region will absorb the light meanwhile white region will reflect it [1].

26.2.3 Smart Cards

Smart card is built with variety of chip with a simple memory consisting of byte of information may have range from 1K up to 64K of microcontroller or multi-application memory. Smart card is used as individual identification; building access and network access are part of a multi-tiered program that is in the final stage of rolling out. The data in smart card can be read when a physical contact is made with a reader [1, 4, 5].

26.3 Methodology

We use the SDLC in completing this project. Among the steps involve here is listed as below:

(a) Project Planning: In the project planning phase, it will present the plan and the structure of the project.
(b) Project Analysis: This section will present the analysis of this project according to the current attendance system and user requirement.
(c) Project Design: This section shows the design and seven main functions will be designed of the project.
(d) Project Implementation: This section will explain which programming language and what tools will be used to develop this project. Monte Carlo simulation and data flow diagram of the project will be presented also.

26.3.1 Project Planning

This project has two parts (shown as Fig. 26.1), one is web page system and the others are smartphone and tablet application.

After student enters or leaves the classroom by scanning RFID card lecturer could view the attendance situation of the class and the movement of different student through web page system, smartphone system and tablet system. Student could view his attendance only and movement only through web page system, smartphone system and tablet system. Besides that, lecturer or student could track the attendance history by both systems.

26.3.2 Project Analysis

Attendance system is one of the most important things to ensure students attend for the particular lectures. Some universities regulate class attendance as compulsory to each student who registered for a particular course. Normally in Malaysian

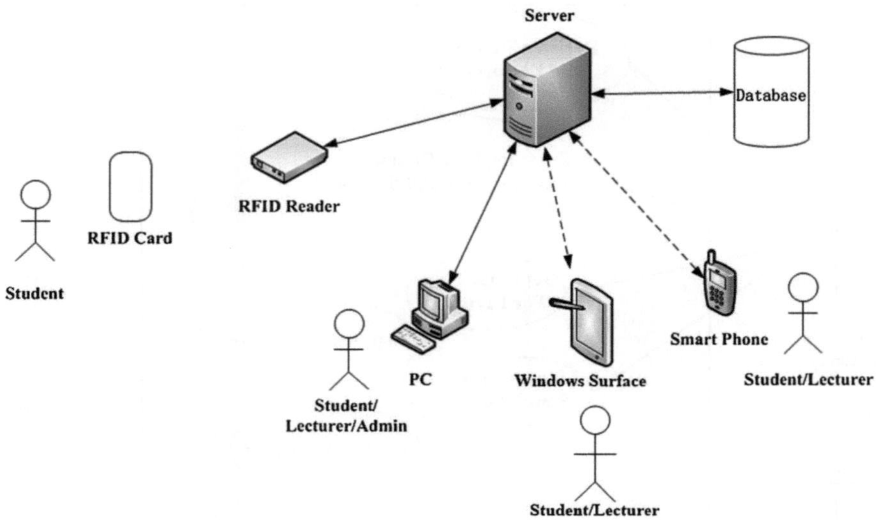

Fig. 26.1 Overall system architecture of smart attendances system

university, students are required to attend the class not less than 80 % per semester otherwise student will be barred from taking any examinations [1].

26.3.3 Project Design

This system has six major functions, those are: It is listed as below:

1. Web-based application of this system can let user to insert, display and track student's attendance and movement in the class.
2. Lecturer could view the situation of his/her courses attendance.
3. Displaying and monitoring lecturer information (courses taught or shared and usage of lecture hall).
4. Administrator can track the allocation of classes.
5. Simulate a classroom using RFID-enabled smart card. Besides that, this system will also have some issues have to be solved like allocating two different courses to one lab at the same time happens. This is due to human error in manually assigning courses according to the classroom. For attendance model, when dozens of students sign attendance, missed reading occurred how students recognize that.

26.3.4 Project Implementation

This project is expected to provide a smart attendance system for different users to sign attendance and view situation of attendance. When the users enter the

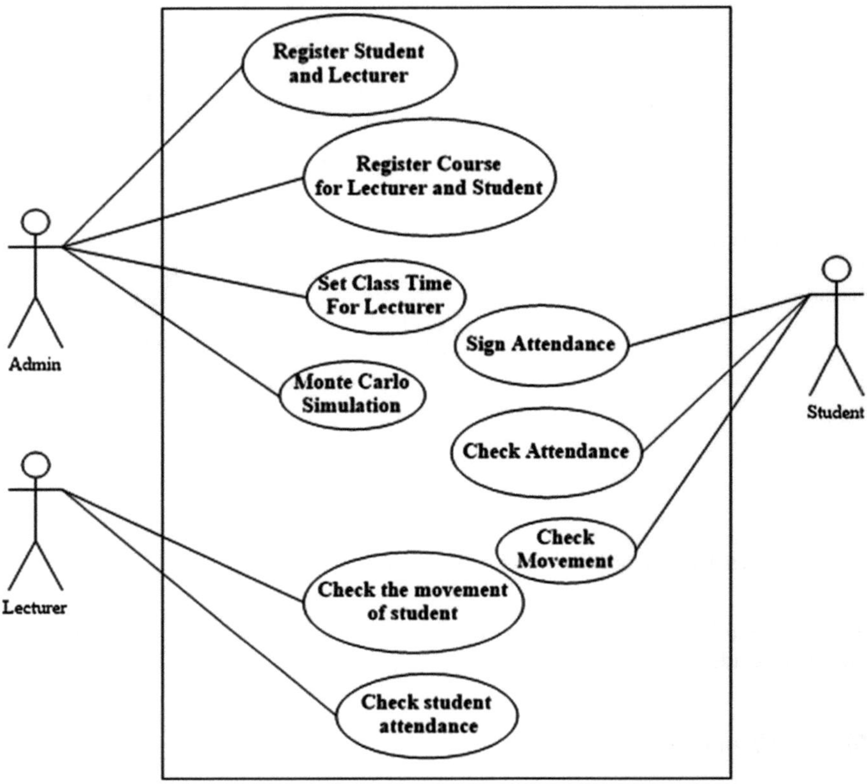

Fig. 26.2 Case diagram of smart attendance system project

classroom or lecture hall, they have the option either to swipe their card on the reader or simply let the card detected by the reader. The card attached with RFID tag, which can be detected by the reader as long as certain range of distance between the tag and the reader is complying. Once the reader detect and obtain the information, it will be then saved to its own database automatically. In addition the lack of automated attendance system in the School of Computer Sciences especially in our lecture halls is our main motivation undesigning this prototype. Figure 26.3 shows the data flow of the whole system.

For web page system, we have used ASP.Net, CSS, and JavaScript through Microsoft Visual Studio 2013 to implement. The other two are Windows Phone application and Windows Surface application, in this application we will adopt Windows Phone 8 platform to implement. The simulation algorithm, which is Monte Carlo, will be integrated as well. Monte Carlo simulation is a method for exploring the sensitivity of a complex system by varying parameters within statistical constraints. These systems can include financial, physical and mathematical models that are simulated in a loop, with statistical uncertainty between simulations.

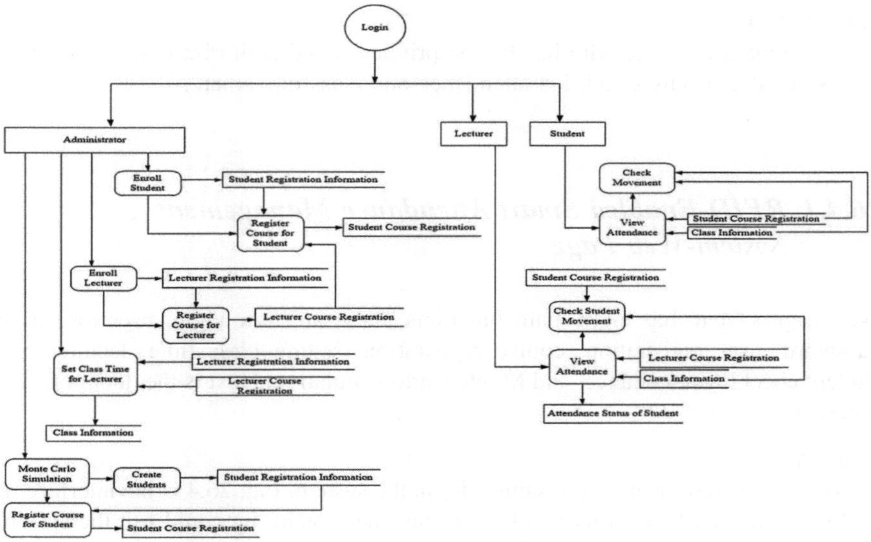

Fig. 26.3 DFD of smart attendances system

The results from the simulation are analyzed to determine the characteristics of the system [2]. Monte Carlo is used in our system aim to generating a new batch of students automatically.

26.4 Implementation

The system includes two different parts. The first part is the web page platform. The second is the ubiquitous platform. All this parts are integrated to function together.

The system is able to function according to three different user roles. This roles and privileges follow the access control and authorization principles. We are following the Discretionary Access Control (DAC) model [6]. Among the three users listed below are:

(a) Administrator
An administrator is a user with the highest privileges and authorization. Typically, an administrator can enroll students and lecturer, register course for students and lecturers, using Monte Carlo simulation.

(b) Lecturer
A lecturer is a user who has enrolled by administrator. Typically, a lecturer can check and track the students' attendance and class movement.

(c) Student
A student is a user who has lowest privileges and authorization. Typically, a student can only check his attendance and class movement.

26.4.1 RFID-Enabled Smart Attendance Management System-Web Page

Web page system has eight main functions; these are user login, user forgetting password, user registration, course registration, setting class time, lecturer and student checking attendance, and Monte Carlo simulation. Next is the detail of each function.

1. Login
   The first step of using his system is login the system, Fig. 26.4 is the interface of "User Login". Once a user is logged into the system, he could use the system according to his roles and rights.
2. Forgetting Password
   If user forgets his/her password user could fetch his/her password through this system. User has to enter the correct email address and matric number or staff number, and then system will generate a six alphabets password and send it to user's mail box. After that, user could login through this new password and change to his/her familiar password.
3. Registration
   After admin click the Enroll Student link system will jump to student registration page. In this page, admin could register student by entering student basic information (First Name, Last Name, Matric No., Email Address, Birthday, Gender, Year of Study, School Name), after all the information finish entering just click Register button, this student will be registered and you will see this student will be record at the bottom table. If one of the information does not enter correctly, system will not register this student and a red "*" will be shown beside the required field. Lecturers cannot register a new account for themselves, and only the administrators have right to register a new account for lecturer. After admin register a new account for lecturer, lecturer could login by

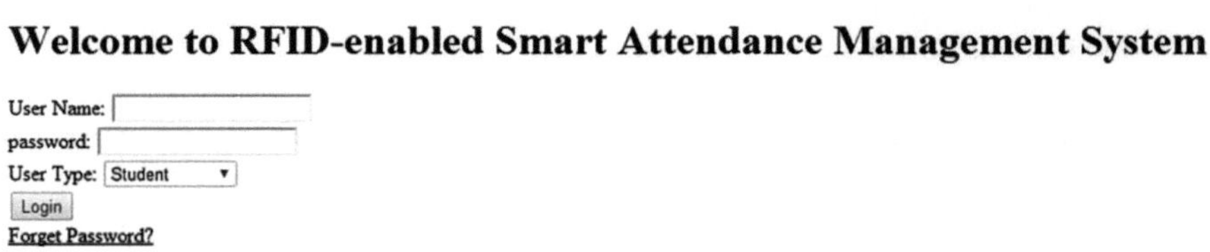

Fig. 26.4 Login interface

his/her email address as username and staff number as initial password. At last but not least, admin could click Finish Registration to go back to home page. After admin click the Enroll Staff link system will jump to staff registration page. In this page, admin could register staff by entering staff basic information (First Name, Last Name, Staff No., Email Address, Birthday, Gender, Year of Study, School Name), after all the information finish entering just click Register button, this staff will be registered and you will see this staff will be record at the bottom table. If one of the information does not enter correctly, system will not register this staff and a red "*" will be shown beside the required field. Staff cannot register a new account for themselves, and only the administrators have right to register a new account for staff. After admin register a new account for staff, staff could login by his/her email address as username and staff number as initial password. At last but not least, admin could click Finish Registration to go back to home page.

4. Register Course
   After admin finishes register a lecturer, admin could help lecturer to register the course. Before the course registration, admin has to choose the teaching year and semester of the course; otherwise the system will not register the course and a red "*" will be shown beside the required field. After admin finishes register a student, admin could help student to register the course. Before the course registration, admin has to choose the teaching year and semester of the course; otherwise the system will invoke the validation checking. If one course is not registered for lecturer, system will not allow any student to register this course also.
5. Admin Set Class Time
   Admin could click the Set Class for Lecturer link to help lecturer to set class time. After admin choose a lecturer and a course the system will jump to the set class time interface, then admin has to choose a building number, room number, start time, end time and class date. Admin could enter a range of class date, and then system will register the class for every week. When admin fill in all the class information, he just click Confirm button, the system will setting class successful, and admin just clicks Finish button to set another course. If admin does not fill in all required information and click Confirm button, the system will not set the class and a red "*" will be shown beside the required field.
6. Lecturer Checking Student Attendance and Movement
   After lecturer login the system, he could click View Attendance link to check student attendance for each class, which he taught. Next lecturer could search the courses by teaching year and semester, after the lecturer click the check button of one course, system will list all the classes, which are set by admin. Next lecturer could click the check button of one class and system will jump to Student Attendance interface. At this time, if it is the first time of checking attendance of this class, lecturer has to click refresh button to get the latest situation of student attendance. If it is not the first time checking, lecturer does not have to click refresh button anymore. And then, system will show the

situation of attendance and attendance statistic of this class. Next, lecturer could click check button of each student to check his/her movement of such class. After lecturer click check button of each student, system will jump to Student Movement interface and it will show the movement record of this student. At the bottom of this interface, system shows the time of this student staying in classroom. At last but not least, lecturer could click Choose Another Student button to check movement of another student.

7. Student Checking Attendance and Movement
   After student login the system, he could click View My Attendance link to check his attendance for each class, which he has to attend. Next student could search the courses by teaching year and semester, after the student click the check button of one course; system will list all the classes and his attendance status. And then, student could click check button of each class and system will jump to Student Movement interface and it will show his movement record. At the bottom of this interface, system shows the time of this student staying in classroom. At last but not least, student could click Choose Another Class button to check his movement of another class.
8. Monte Carlo Simulation
   After admin login the system, admin could click Monte Carlo Simulation link to generate a new batch of students and register the courses for each through the simulator. When the system jump to the Monte Carlo Simulation interface, click Create Student button, and system will generate a batch of students (200 students) and admin could click Log button to check the students' information. After that, admin could click Register Course button to register the courses for each student, which are generated by Monte Carlo simulator. Then, admin could click Register Course button of each student to check the situation of course registration. Figure 26.5 is the simple code of Monte Carlo Simulation.

26.4.2 Ubiquitous Platform

In this section we will introduce the ubiquitous characteristics of this system. For this system we have embedded RFID model, smartphone and tablet. Next we will introduce the detail of each part.

1. RFID
   In RFID platform we use MIFARE522_MODULE as hardware reader and passive tag to implement the function. This platform has three main functions; these are student signing attendance, duplication reading, miss reading, and sending email, Fig. 26.6 shows the RFID model and tags. Next is the detail of each function.

Fig. 26.5 Monte Carlo simple code

```
public string generateEmailAddress()
{
        ...
}
DateTime RandomDay()
{
        ...
}
public string generateTime()
{
        ...
}
public static string GenRandomLastName()
{
        ...
}
public static string GenRandomFirstName()
{
        ...
}
```

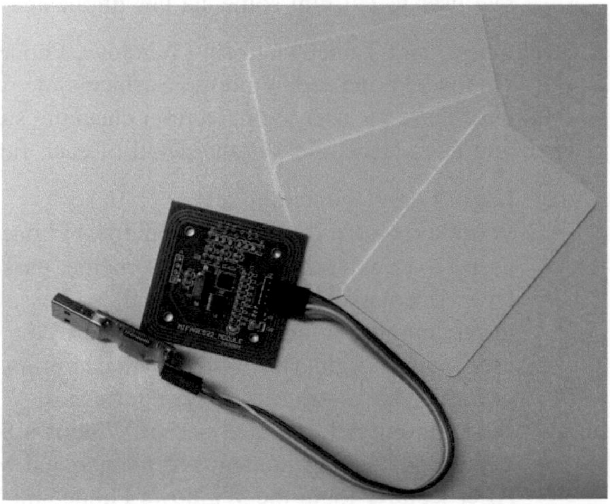

Fig. 26.6 RFID reader and tags

(a) Signing Attendance for Students
When students enter the classroom he just let his RFID card on the reader and he will hear two beep sounds, it means the reader have detected the card and recorded student enter the classroom. If students want to go out of the class, he also has to scan the card and he will hear a beep sound.

It means the system have record the students already out of the classroom. At last, lecturer will calculate the attendance time and determine the attendance for each student through this system.

(b) RFID-Attendance event Duplication Reading
The aim of doing a duplication reading is in case of someone duplicate the card and sign attendance for students. After the students scan the card, system will check the truth of the card, if the card is a fake one, system will beep three times and it will not be record in the system. If the card is a real one, system will record the data in the system.

(c) RFID-Attendance event Miss Reading
The aim of miss reading is in case of students just put his card on the reader but the system will not record this student and students do not know. After a student scans his card, the system will give a feedback by beep sound and send an email to student. When students do not hear the beep sound and do not receive the email, it means system did not detect the card and students have to scan again.

(d) Email Generation
In this function, system will send an email to user in two situations. One is when students scan the card, system will send an email to students and tell them they current status. Second, if a person uses a fake card to sign attendance for a student, system will find it out and send an email to the real user to tell him someone has duplicated your card.

2. Smartphone and Tablet Platform (Windows Phone and Windows Surface)
For Windows Phone and Windows Surface systems, each of one has three main functions; these are user login, lecturer checking student attendance, and student checking attendance. Next is the detail of each function.

(a) Login/Logoff Function
When user login the system (Windows Phone and Windows Surface part), user has to enter their correct username, password and choose correct use type. Username is the users email address, and initial password is user matric number or staff number. If one of the information does not correct user cannot login and system will show an error message to user. Figure 26.7 shows the user login interface of Windows Phone and Fig. 26.8 shows the user login interface of Windows Surface.

(b) Lecturer Checking Student Attendance and Movement
After lecturer login the Windows Phone and Windows Surface system, he could press View Attendance button to check student attendance for each class, which he taught. Next the system will show the course list base on teaching year, after the lecturer press one of the courses, system will list all the classes, which are set by admin. Next lecturer could press one of the classes and system will jump to Student Attendance interface. System will show the situation of attendance and if lecturer presses the Statistic button, system will jump to next interface and show the attendance statistic of this

Fig. 26.7 User login (Windows Phone)

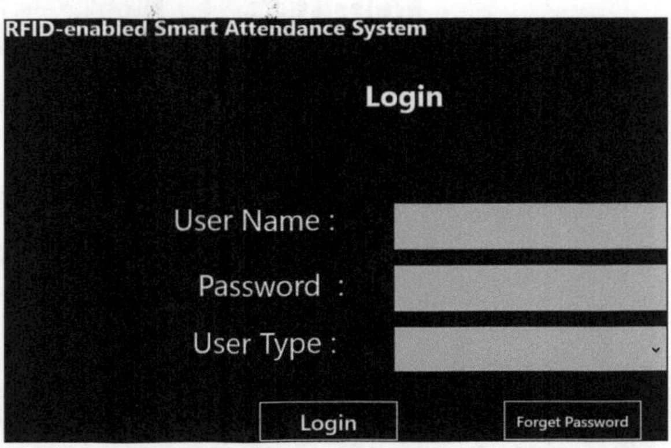

Fig. 26.8 User login (Windows Surface)

class. Next, lecturer could press any of the students to check his/her movement of this class. After lecturer press one of the students, system will jump to Student Movement interface and it will show the movement record of this student. At the bottom of this interface, system shows the time of this student staying in classroom. At last but not least, lecturer could press back button to check movement of another student. Figures 26.9 and 26.10 show the interface of lecturers check student attendance.

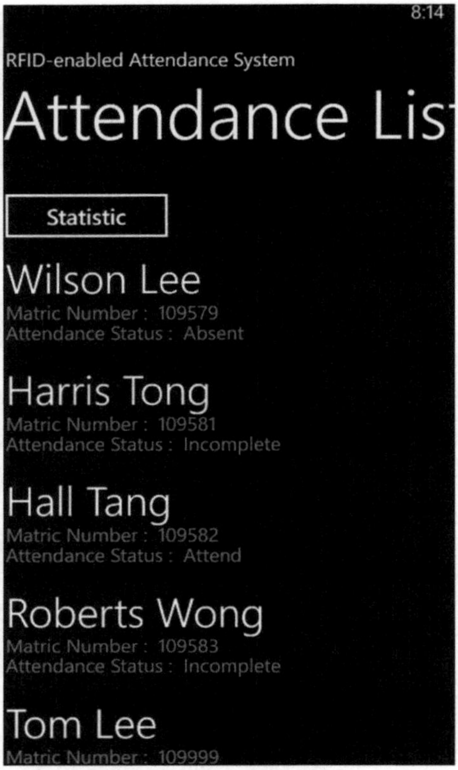

Fig. 26.9 Lecturers check students attendance (Windows Phone)

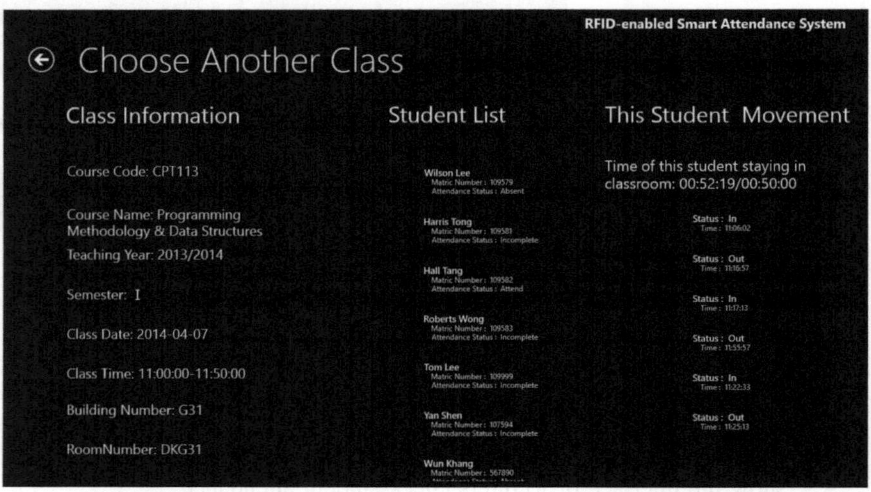

Fig. 26.10 Lecturers check students attendance (Windows Surface)

(c) Student Checking Attendance and Movement
After student login the smartphone system or tablet system, he could press View Attendance button to check his attendance for each class. Next the system will show the course list base on teaching year, after the student press one of the courses, system will list all the classes, and his attendance status. And then, student could press one of the classes and system will jump to Student Movement interface and it will show his movement record. At the bottom of this interface, system shows the time of this student staying in classroom. At last but not least, student could press back button to check his movement of another class. Figures 26.11 and 26.12 show the interface of students check student attendance.

26.5 Testing

In order to do the testing, users log in as the administrator, lecturer and students. RFID system, web page system, smartphone application and tablet application were fully tested. Below are the results of the function testing.

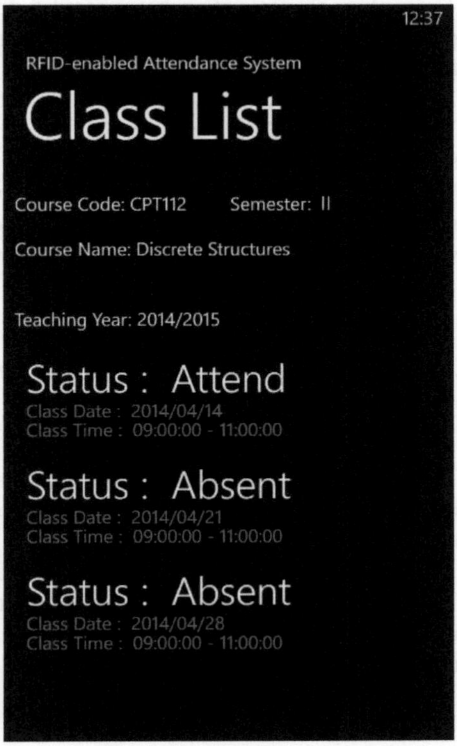

Fig. 26.11 Students check students attendance (Windows Phone)

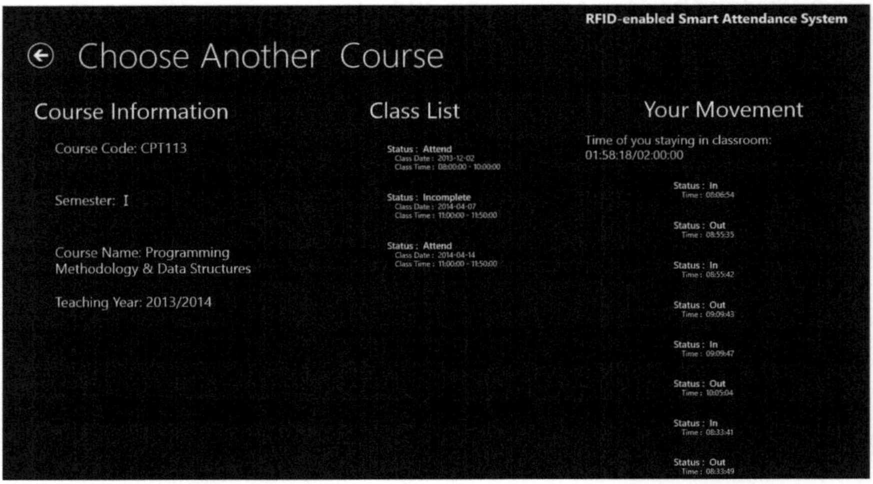

Fig. 26.12 Students check students attendance (Windows Surface)

26.5.1 Testing Web Page Function

The web page application has several functions such as user login, user changing password, lecturer checking student attendance and etc. Below are the tested functions.

Based on Table 26.1, we demonstrate that all the functions work successfully. The role of the administrator could register student and lecturer, register course for student and lecturer, setting class time for lecturer and simulate a batch of student by Monte Carlo simulator. The role of lecturer could check student attendance and movement of each student and check attendance statistic of each class. The role of student could check his own attendance and movement.

26.5.2 Testing Windows Phone and Windows Surface Function

The smartphone and tablet application has several functions such as user login, user changing password, lecturer and student checking attendance and movement and etc. Below are the tested functions.

Based on Table 26.2, we demonstrate that all the functions work successfully. The role of lecturer could check student attendance and movement of each student and check attendance statistic of each class. The role of student could check his own attendance and movement.

Table 26.1 Web page function testing

No.	Functions	Tested and working well
1	User login	√
2	User forgetting password	√
3	User changing password	√
4	Admin register student	√
5	Admin register lecturer	√
6	Admin register course for student	√
7	Admin register course for lecturer	√
8	Admin setting class time	√
9	Mote Carlo simulation	√
10	Lecturer checking student attendance	√
11	Lecturer checking student movement	√
12	Class statistic	√
13	Student checking attendance	√
14	Student checking movement	√
15	Sending email	√

Table 26.2 Windows Phone and Windows Surface function testing

No.	Function	Tested and working well
1	User login	√
2	User forgetting password	√
3	User changing password	√
4	Lecturer checking student attendance	√
5	Class statistic	√
6	Lecturer checking student movement	√
7	Student checking attendance	√
8	Student checking movement	√
9	Sending email	√

26.5.3 Testing RFID Function

In RFID part, it has several functions such as students signing attendance, duplication reading checking, miss reading feedback and sending email. Below are the tested functions.

Based on Table 26.3, we demonstrate that all functions work successfully. RFID system could perform students signing attendance, duplication reading checking, miss reading feedback and email generation.

Table 26.3 RFID function testing

No.	Function	Tested and working well
1	Signing attendance	√
2	Duplication reading	√
3	Miss reading	√
4	Sending email	√

26.6 Discussion

This RFID smart attendance system implemented based on the requirement that has been set during the analysis stage and able to expand the requirement during the design stage to make the attendance system more advances.

However, this project still has some challenges and limitations to be improved. There are several weakness of this system have to be improved. First, the connection port of RFID reader is TTL port and the connection port of server uses USB port, but the efficient transmission distance between these two ports is about 500 m, if the user wants to extend the distance he has to use repeater, this is very troublesome. Second, in the process of students scanning card and entering classroom, sometime the system will send the email to students, and in the period of system sending email, reader cannot read the card until sending finished. This phenomenon decreases the efficiency of reading cards. Third, if a student scans his card at the beginning of this class, and then he leaves the classroom until the class finished he comes back and scans his card. And student could pass the card to other student to let them help to sign attendance, for these phenomena system cannot be detected. So that, this system still cannot eliminate student absent the class completely. In the future, this system could be combined with Near Field Communication (NFC) function to improve the weakness of this system. Nowadays, NFC becomes more and more popular in mobile device, and this function is based on RFID technology, hence this system could be combined with this function and let students to sign attendance by using smartphone and tablet. At last but not least, with the development of technology the completeness of this system could be improved and this system could be more and more completed.

26.7 Conclusion

This project provides a smart attendance system for students to sign attendance and view movement of student. When the students enter the classroom or lecture hall, they have to let the card detected by the reader. The reader can detect the card attached with RFID tag as long as certain range of distance between the tag and the reader is complying. Once the reader detect and obtain the information, it will be then saved to its own database automatically. Besides that, students can leave the classroom but they have to let the card detected by the reader also. So that teachers or lecturers not only can view the percentage of coming students, earl leave and absent students, but also they can see which students come, early leave and absent no matter by web page system, smartphone system or tablet system. As a student, he/she could through web page system, smartphone system or tablet system to look up his/her attendance situation and movement situation. The outcome of the project especially the importance of RFID based attendance system can stand as an extension to the existing student card currently used in our University. In addition the lack of automated attendance system in the School of Computer Sciences especially in our lecture halls is our main motivation undesigning this system.

Although this project has met the basic requirement, it still has space to be improved. In the future this project could combine with the NFC technology, the RFID reader and tag could be change to a mobile device has NFC function. Then students could use only smartphone to sign attendance and it is very convenience.

Acknowledgments The authors would like to extend our greatest appreciation and gratitude to the organization of Innovate Malaysia Design Competition and Microsoft for providing various technical and soft skills training, technical support as well as the Nokia smartphone.

References

1. Abd Wahab MH, Abdul Kadir H, Mohd Yusof MN, Sanudin R, Tomari MR (2009) Class attendance system using active RFID: a review. Compilation of papers 2009, vol 2, pp 9–21
2. Arulogun OT, Olatunbosun A, Fakolujo OA, Olaniyi OM (2013) RFID-based students attendance management system. Int J Sci Eng Res 4(2):1–9. ISSN 2229-5518
3. Jiang L, Li Y (2012) Attendance technology on campus research and application. In: National conference on information technology and computer science (CITCS 2012), pp 40–42
4. Kassim M, Mazlan H, Zaini N, Salleh MK (2012) Web-based student attendance system using RFID. In: Control and system graduate research colloquium (ICSGRC), 2012 IEEE, pp 213–218
5. Lim TS, Sim SC, Mansor MM (2009) RFID based attendance system. In: Industrial electronics and applications (ISIEA), 2009 IEEE, pp 778–782
6. Stalling W, Bauer M, Howard M (2011) Computer security: principles and practice. Pearson Education Limited. ISBN No. 9780132775069
7. Ujan EIA, Ismaili IA (2011) Biometric attendance system. In: 2011 IEEE/ICME international conference on complex medical engineering (CME), pp 499–501

Chapter 27
Performance Evaluation of IEEE 802.15.6 MAC with User Priorities for Medical Applications

Li Yang, Changle Li, Yueyang Song, Xiaoming Yuan and Yanle Lei

Abstract In order to satisfy the heterogeneous service requirements from different applications and the complex channel characters owing to body motions, IEEE 802.15.6 standard was established as a new solution for Wireless Body Area Networks (WBANs). In this paper, we evaluate the effect of user priorities (UPs) on the performance of IEEE 802.15.6 CSMA/CA channel access mechanism in narrow band. Simulation metrics mainly focus on the normalized throughput and average packet delay in which the traffic arrival rate and traffic distribution vary. In addition, we make a comparison with the non-priority CSMA/CA in performance which concludes that the IEEE 802.15.6 with user priorities performs better in specific situation.

Keywords WBAN · IEEE 802.15.6 · CSMA/CA · User priorities

27.1 Introduction

Wireless Body Area Network (WBAN) is a kind of short-range, wireless communication networks, aiming at providing the access services of low-power, high reliability and low latency, which can be exploited in many fields that varies from the medical application (e.g. health monitoring), the consumer electronics to the personal entertainment. Therefore, the broad prospects have motivated many researches on the key technologies and standardization process of WBANs.

As a solution for Wireless Personal Area Network (WPAN), IEEE 802.15.4 [5] provides technologies to operate around human body, but due to the effect of human body on the radio channel and the diversified demands in different applications, it suffers restrictions when used in the timely and reliable situation. Therefore, IEEE 802.15.6 [6] was released as the appropriate standard for short-range, wireless communication in the vicinity of, or inside human body.

L. Yang · C. Li (✉) · Y. Song · X. Yuan · Y. Lei
State Key Laboratory of Integrated Services Networks, Xidian University, Xi'an 710071, Shaanxi, China
e-mail: clli@mail.xidian.edu.cn

J.J. (Jong Hyuk) Park et al. (eds.), *Future Information Technology - II*,
Lecture Notes in Electrical Engineering 329,
DOI 10.1007/978-94-017-9558-6_27

Compared with IEEE 802.15.4, several improvements have been made by IEEE 802.15.6 to provide high quality of service (QoS) and extreme high power efficiency. IEEE 802.15.6 designs three different Physical Layers (PHYs) to adapt the broad range of possible applications and constructs a more flexible frame structure that supports multiple access modes. However, the greatest difference is that assigning different priorities based on the traffic type, which guarantees the high-priority data, for example, the emergency data, transmit timely and reliably.

Most of existing researches pay attention to Medium Access Control (MAC) of IEEE 802.15.6. Reference [3] draws an overview of WBAN main applications, technologies and summarizes the advantages of IEEE 802.15.6, while [2, 8] give detailed descriptions of MAC functionalities in IEEE 802.15.6. Reference [12] presents the theoretical maximum throughput and minimum delay limits of IEEE 802.15.6 under different frequency bands and data rates. Reference [7] evaluates the performance of IEEE 802.15.6 MAC in terms of energy consumption and energy efficiency.

There are also some literatures that study the effect of user priorities on the performance of IEEE 802.15.6. Reference [1] analyzes the performance of IEEE 802.15.6 MAC where there are three user priorities (UP1, UP3 and UP5) in the whole network, which is not rigorous due to ignoring the other UPs especially the emergency medical traffic of UP6 and UP7. Reference [9] presents a three-dimensional Markov model to evaluate the performance of CSMA/CA scheme in saturation condition, which means there is always a packet in the queue waiting to transmit, thus it can't be used in the medical application directly. Reference [11] investigates the effectiveness of user priorities in IEEE 802.15.6 with focusing on the influence of the access phase lengths, but the model of arrival traffic is Bernoulli arrival process, which deviates from the real medical scenarios because the primary type of traffic is periodic type traffic. Compared with the previous works, we evaluate the effect of user priorities on the performance of IEEE 802.15.6 CSMA/CA versus the traffic load and traffic distribution in narrow band. Both the unsaturated and saturated situations are taken into account, periodically arrival process is considered as the model of arrival traffic, which is more suitable for the real medical scenario, such as the continuous monitoring of the human body. In addition, we also make a comparison with the non-priority CSMA/CA.

The remainder of the paper is organized as follows: Sect. 27.2 briefly describes IEEE 802.15.6 MAC with the details of CSMA/CA. In Sect. 27.3 simulation models, parameters and results are presented. According to the simulation results, Sect. 27.4 concludes the paper and makes an outlook on future work.

27.2 Overview of IEEE 802.15.6 MAC

IEEE 802.15.6 supports three access modes. Here we focus on the beacon mode with superframes that the hub transmits a beacon at the start of each superframe to inform the information related to BAN identification, synchronization, and superframe structure.

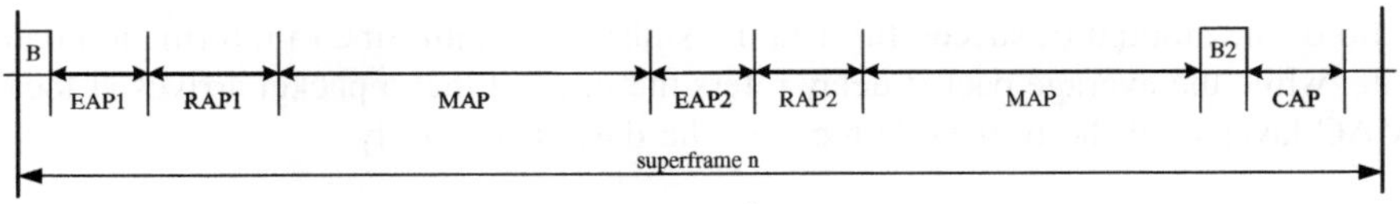

Fig. 27.1 Superframe structure of the beacon mode with superframes

Table 27.1 CW bounds and UP mapping for CSMA/CA

Priority	UP	Traffic designation	CWmin	CWmax
Lowest	0	Background (BK)	16	64
	1	Best effort (BE)	16	32
	2	Excellent effort (EE)	8	32
	3	Video (VI)	8	16
	4	Voice (VO)	4	16
	5	Medical data or network control	4	8
	6	High-priority medical data or network control	2	8
Highest	7	Emergency or medical implant event report	1	4

Figure 27.1 shows the superframe structure under beacon mode with superframes. Among the seven periods, EAP1, EAP2, RAP1, RAP2 and CAP access phases are used for the contended allocation based on CSMA/CA or slotted Aloha scheme while the MAP is composed of uplink allocation intervals, downlink allocation intervals and bilink allocation intervals.

Depending on the characteristics of different traffic flows, eight user priorities (UPs) which span from UP0 to UP7 are assigned to distinguish them. The higher priorities represent urgent traffic while the lowers are for general traffic. The predefined relationship between UP and contention window (CW) bounds, CWmin and CWmax, for CSMA/CA are showed in Table 27.1.

The details of CSMA/CA scheme are described as follows: When data arrives, the node firstly selects a random integer uniformly distributed over the interval [1, CW] as the back-off counter value. Then the node will perform the clear channel assessment (CCA) detection, the back-off counter subtracts by one in the unlock state. Once the back-off counter decreases to zero and the remaining time is long enough, it transmits the data immediately. If this transmission failed, the CW and back-off counter change their stage based on the corresponding rules in [6] until the packet is transmitted successfully or has reached the max retransmission times.

27.3 Performance Evaluation

In this section, we make simulation on the platform MIRAI-SF [10]. The Two performance metrics considered in our work are the normalized throughput and average packet delay, respectively. The normalized throughput is defined as the ratio of the amount of successful data transmission in unit time to information data rate, while the average packet delay refers the interval that a packet arrives at local MAC layer until the peer node receives the data successfully.

27.3.1 Simulation Scenario

A one-hop star network is chosen as the network topology which consists of a hub and eight nodes, as shown in Fig. 27.2. Each node generates a periodic data flow that represents a kind of UPs varied from UP0 to UP7, competes for allocation slot by CSMA/CA. Considering the medical sensing where most data transmissions are initiated by the sensor nodes, the downlink traffic is not considered in this paper.

The key parameters we used in the simulation are listed in Table 27.2.

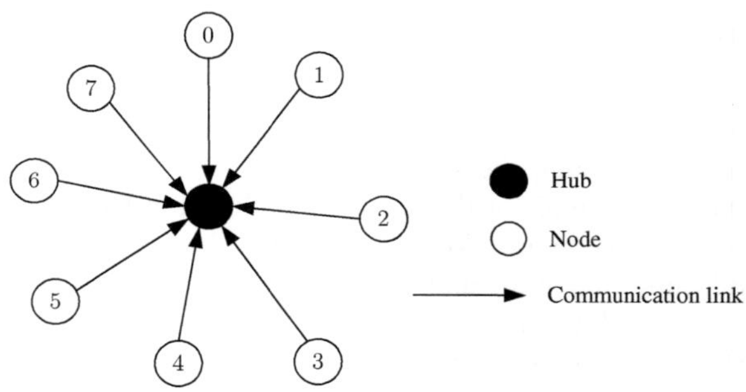

Fig. 27.2 The network topology used in the simulation

Table 27.2 Simulation parameters

Parameters	Default value
IEEE 802.15.6 CSMA/CA	
Frequency band	2,400–2,483.5 MHz
Information data rate	971.4 kbps
Payload size	255 Bytes
Symbol rate	600 ksps
MAC header	9 Bytes
PLCP header	31 bits
PLCP preamble	90 bits
ACK	193 bits
pSIFS	75 μs
CSMA/CA slot time	145 μs
Non-priority CSMA/CA	
CWmin	3
CWmax	5

27.3.2 Simulation Results

In this part, we evaluate the performance of IEEE 802.15.6 CSMA/CA for different UPs versus the arrival rate of packets and make a comparison with non-priority CSMA/CA where all nodes have fair chances to access channel.

From Figs. 27.3 and 27.4, one can see that large difference exists on the performance of nodes with different UPs. With the arrival rate of traffic in the network increasing, the normalized throughput of UP7 and UP6 can achieve improvement constantly, whereas the other UPs increase firstly but then fall down. The average packet delay of each UP is almost constant at first, then increases rapidly, and finally flattens out, whereas the UP7 remains almost unchanged. One reason is when the arrival rate is low, each packet can access channel effectively, but after getting a certain value, fierce collisions lead the network to be saturated. The other is the different CW values for different UPs, the higher UPs access channel

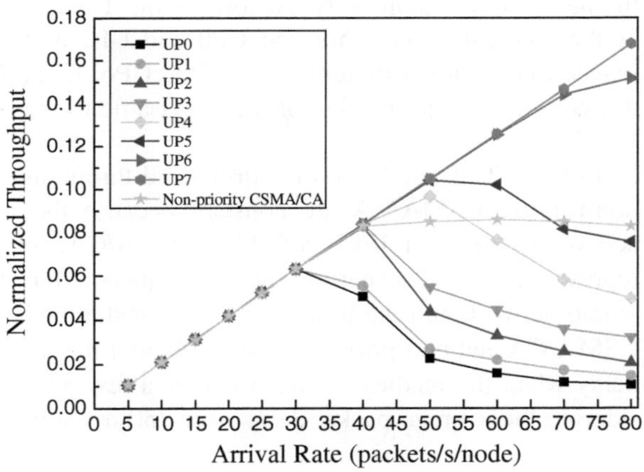

Fig. 27.3 Comparison of normalized throughput for all UPs in IEEE 802.15.6 CSMA/CA and non-priority CSMA/CA

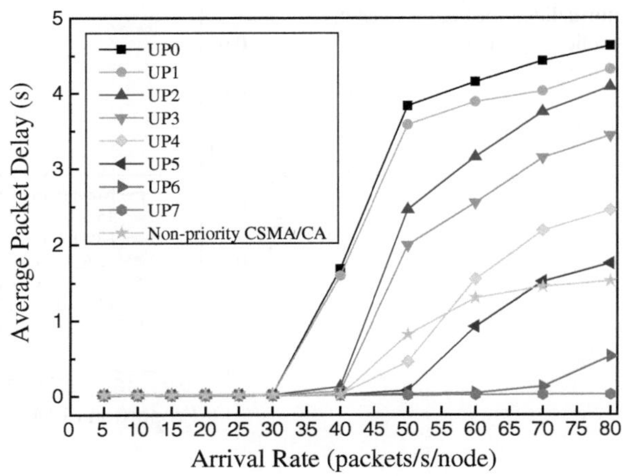

Fig. 27.4 Comparison of average packet delay of all UPs in IEEE 802.15.6 CSMA/CA and non-priority CSMA/CA

frequently as the lower CWmin and CWmax, while the lower UPs may make more backoffs before accessing channel. The higher of UP can obtain the higher maximum throughput and lower average packet delay simultaneously.

Figures 27.3 and 27.4 also show that the non-priority CSMA/CA has the approximate performance with IEEE 802.15.6 CSMA/CA when the arrival rate is below 30 packets/s. From then it performs better than IEEE 802.15.6 CSMA/CA for UP0, UP1 firstly, better than the UP2 and UP3, UP4, UP5 at the arrival rate of 40, 55 and 70 packets/s, respectively. Nodes in the non-priority CSMA/CA access channel in fair, whereas the higher UPs occupy the channel frequently such that gives rise to the collision with other UPs in the IEEE 802.15.6 CSMA/CA. After the arrival rate has reached a relatively high value, the UP6 and UP7 still perform better than the non-priority CSMA/CA, which indicates the IEEE 802.15.6 CSMA/CA can guarantee higher reliability and lower delay for the high UPs, especially the emergency data and high-priority medical data.

In this part, we evaluate the performance of IEEE 802.15.6 CSMA/CA varies with the traffic distribution in the whole network. Here we assume only two UPs in the network simultaneously. According the Table 27.1, we treat the UP6 and UP7 as the medical traffic while the UP0 and UP2 as the non-medical traffic. Performance under three different cases-UP0&UP6, UP2&UP6 and UP2&UP7 are conducted when the proportion of medical traffic to the overall traffic varies from 0 to 100 %.

In Figs. 27.5 and 27.6, the normalized throughput and average packet delay in non-priority CSMA/CA are constant because the nodes have equal chances to access channel. For IEEE 802.15.6 CSMA/CA, the two curves with same color represent a kind of scenarios where the square is for the average performance of the whole network and the triangle is only for the medical traffic. The IEEE 802.15.6 CSMA/CA and non-priority CSMA/CA offer the same throughput and approximate delay when the medical traffic occupies a few part of traffic in the network. Considering the average packet delay over 250 ms is unacceptable for general WBAN

Fig. 27.5 Comparison of normalized throughput versus traffic distribution

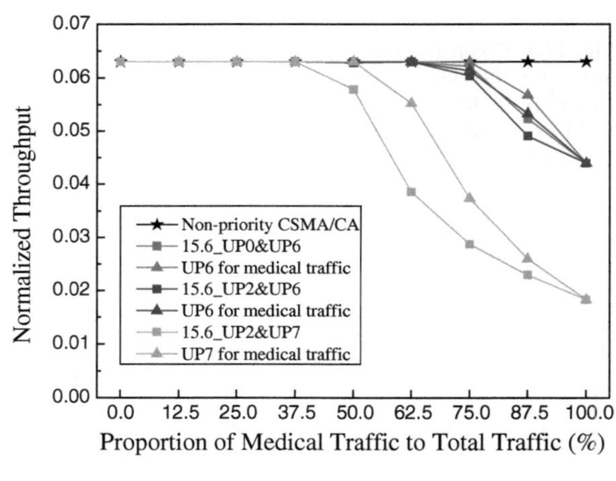

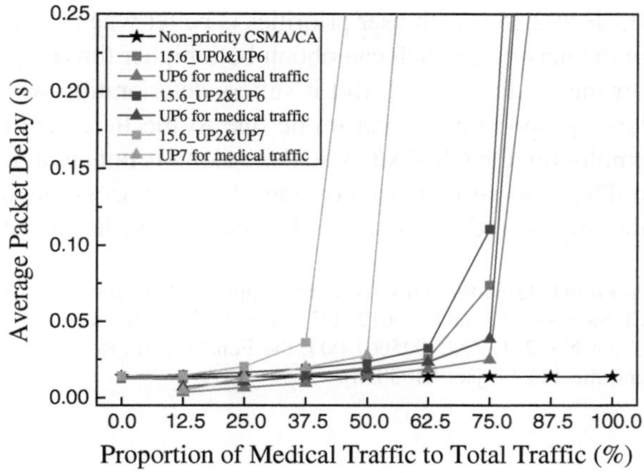

Fig. 27.6 Comparison of average packet delay versus traffic distribution

applications [4], Fig. 27.6 only shows the performance of IEEE 802.15.6 CSMA/CA in the acceptable range of delay. It should be noted that the IEEE 802.15.6 CSMA/CA guarantees a lower average delay for the medical traffic when there is few medical traffic. The reduction of delay for medical traffic is up to 74.5 % in the case of UP0&UP6 when the proportion of medical traffic is 12.5 %, which indicates that employing the IEEE 802.15.6 CSMA/CA is better because the latency requirement of a medical data is much lower than a background data.

However, after the occupancy of medical traffic exceeds a certain value, the IEEE 802.15.6 CSMA/CA will suffer deterioration and perform poorer than the non-priority CSMA/CA due to the frequent collisions. For the medical traffic, the corresponding contention window is so small that the probability to choose same back-off counter increases, which always leads collisions in the first few transmission times. Even so, both the throughput and delay for medical traffic are still well above the average level.

Compared with the performance in the case of UP2&UP6, the one in UP2&UP7 is much worse because the higher UPs for medical traffic have smaller CW to contend for channel frequently, which is more selfish to complete its data transmission at the expense of hindering transmission of lower UPs and of increasing the internal competition among higher UPs. However, the one in UP0&UP6 has better performance because the lower UPs have larger CW to avoid conflicting with higher UPs, thus improves the performance of the whole network by alleviating the congestion of the network.

27.4 Conclusion

For the IEEE 802.15.6 CSMA/CA, user priorities mechanism makes a big influence on network performance, it guarantees reliable and timely service for higher UPs with sacrificing the performance of lower UPs. What's more, the performance of

IEEE 802.15.6 with user priorities is better when there is lightweight medical traffic in the network, which can obtain both higher throughput and lower delay especially for the medical traffic. But it suffers deterioration when the medical traffic holds a large proportion of total traffic due to the fierce collisions. It also concludes that employing the CSMA/CA scheme alone is not enough to meet all needs of medical traffic, thus we will also consider the other access modes in IEEE 802.15.6 such as the improvised access and scheduled access in the future.

Acknowledgments This work was supported by the National Natural Science Foundation of China under Grant No. 61271176, the National Science and Technology Major Project under Grant No. 2013ZX03005007-003, the Fundamental Research Funds for the Central Universities, and the 111 Project (B08038).

References

1. An N, Wang P, Yi C, Li Y (2013) Performance analysis of CSMA/CA based on the IEEE 802.15.6 MAC protocol. In: 15th IEEE international conference on communication technology (ICCT), China, pp 539–544
2. Brada N, Belhaj S, Chaari L, Kamoun L (2011) Study of medium access mechanisms under IEEE 802.15.6 standard. In: 4th Joint IFIP on wireless and mobile networking conference (WMNC), Toulouse, pp 1–6
3. Cavallari R, Martelli F, Rosini R, Buratti C (2014) A survey on wireless body area networks: technologies and design challenges. IEEE Communications Surveys & Tutorials 16(3): 1635–1657
4. Cordeiro C (2007) Use cases, applications, and requirements for bans. Doc.: 802.15-07-0564
5. IEEE 802.15.4 Standard, Part 15.4: wireless medium access control (MAC) and physical layer (PHY) specifications for low-rate wireless personal area networks (WPANs) (2003)
6. IEEE Standard for local and metropolitan area networks, Part 15.6: wireless body area networks (2012)
7. Jung BH, Akbar RU, Sung DK (2012) Throughput, energy consumption, and energy efficiency of IEEE 802.15.6 body area network (BAN) MAC protocol. In: 23rd IEEE international symposium on personal indoor and mobile radio communications (PIMRC), Sydney, pp 584–589
8. Kwak KS, Ullah S, Ullah N (2010) An overview of IEEE 802.15.6 standard. In: 3rd IEEE international symposium on applied sciences in biomedical and communication technologies (ISABEL), Rome, pp 1–6
9. Li C, Geng X, Yuan J, Sun T (2013) Performance analysis of IEEE 802.15.6 MAC protocol in beacon mode with superframes. KSII Transactions on Internet and Information Systems 7(5): 1108–1130
10. MIRAI-SF Simulator 4.0 User Manual. http://www2.nict.go.jp/w/w122/old/mt/b192/mirai-sf/overview_e.html
11. Rashwand S, Mišic J, Mišic VB Analysis of CSMA/CA mechanism of IEEE 802.15.6 under non-saturation regime, to appear in IEEE Transactions on Parallel and Distributed Systems
12. Ullah S, Chen M, Kwak KS (2012) Throughput and delay analysis of IEEE 802.15.6-based CSMA/CA protocol. J Med Syst 36:3875–3891

Chapter 28
Implementation of Policy-Based Personalized Wellness Service in IoT Environment

Namje Park

Abstract In this paper, we describe the security analysis and implementation leveraging globally networked mobile RFID-based wellness service in IoT Environment. We propose a secure mobile RFID-based wellness service framework leveraging mobile networking. The framework provides a means for safe use of mobile phone-based RFID services by providing security to personalized wellness service.

28.1 Introduction

The typical architecture of an Radio frequency identification (RFID) system, as defined by EPCglobal [1], comprises tags embedded or attached to objects, tag readers that read tag information, and a backend Information Services (IS) server that provides the required information. The tag reader can be designed to be portable or handheld, which allows for several possible applications. While RFID is most commonly used in business-to-business (B2B) commerce for managing supply channels, distribution, and logistics, there is also growing interest in the integration of tag readers with mobile phones, allowing individuals to collect and use tag data, such as in business-to-customer (B2C) marketing. And although current implementations have been limited to movie promotions and museums, where information security is not a major concern, continued development will see more frequent adoption in such fields as retail, medical care, and electrical drafts, where security and privacy are indispensable.

Researchers have developed many techniques to address the various security flaws in RFID systems [2, 5, 6], the simplest being to "kill" tags before they are in the hands of the user [10]. However, because these low-cost tags have numerous

N. Park (✉)
Department of Computer Education, Teachers College, Jeju National University,
61 Iljudong-ro, Jeju-si, Jeju 690-781, South Korea
e-mail: namjepark@jejunu.ac.kr

J.J. (Jong Hyuk) Park et al. (eds.), *Future Information Technology - II*,
Lecture Notes in Electrical Engineering 329,
DOI 10.1007/978-94-017-9558-6_28

applications, the user may want the tag to remain active. As one solution, Rieback et al. proposed the *RFID Guardian* system [10], which relies on a strong proxy device—a mobile phone or PDA—to mediate access by an external reader to tags for auditing of scans and tags, key management, access control, and user authentication. The *RFID Enhancer Proxy* (REP) proposed by Juels et al. requires the use of a similar higher-computing-power proxy device [5] but provides better tag acquisition and ownership transfer. Kim et al.'s *Mobile Agent for RFID Privacy Protection* (MARP) is a further development that aims to provide high-level privacy [6] by employing a public key center that manages keys for the readers, tags, server, and proxy. Kim et al. also recently proposed a scheme suitable for mobile-phone-based reader systems [7], but the method only provides reader-based authentication, which is not sufficient.

Here we propose a secure framework for mobile-phone based RFID services using personal privacy-policy-based access control for personalized ultra-high frequency (UHF) tags employing the Electronic Product Code (EPC). The framework, called M-RPS, has dynamic capabilities that extend upon extent trust-building service mechanisms for RFID systems. This new technology aims to provide absolute confidentiality with only basic tags.

28.2 Strategic Security Framework Architecture

28.2.1 Multilateral Approaches for Improved Privacy

This technology is aimed at RFID application services like authentication of tag, reader, and owner, privacy protection, and non-traceable payment system where stricter security is needed.

- Approach of Platform Level: This technology for information portal service security in offering various mobile RFID applications consists of application portal gateway, information service server, terminal security application, payment server, and privacy protection server and provides a combined environment to build a mobile RFID security application service easily.
- Approach of Protocol Level: It assists write and kill passwords provided by Electronic Product Code (EPC) Class1 Gen2 for mobile RFID tag/reader and uses a recording technology preventing tag tracking. Information technology solves security vulnerability in mobile RFID terminals that accept WIPI as middleware in the mobile RFID reader/application part and provides End-to-End (E2E) security solutions from the RFID reader to its applications through WIPI based mobile RFID terminal security/code treatment modules.
- Approach of Privacy Level: This technology is intended to solve the infringement of privacy, or random acquisition of personal information by those with RFID readers from those with RFID attached objects in the mobile RFID circumstance except when taking place in companies or retail shops that try to collect personal

information. The main assumptions are privacy in the mobile RFID circumstance when a person holds a tag attached object and both information on his/her personal identity (reference number, name, etc.) and the tag's information of the commodity are connected. Owners have the option to allow access to any personal information on the object's tag by authorized persons like a pharmacist or doctor but limit or completely restrict access to unauthorized persons.

28.3 Implementation and Wellness Service

28.3.1 Implementation of Customized Wellness Service

We implemented the proposed system for tracking patient care at a hospital. Context-relevant information is important in a ubiquitous computing environment for providing medical care. Different user policies are necessary for patient tags and product tags in EPCglobal's enterprise application. This ubiquitous sharing system for medical information poses a serious threat to the privacy of personal medical information such location, health, and clinical history. Standards such as Health Level Seven (HL7) do not allow customization and do not include rigorous privacy mechanisms. Therefore, we propose a mechanism that manages privacy policy in a user-centric manner for ubiquitous medical care. It is flexible, secure, and can be integrated with a cryptographic algorithm for mitigating the aforementioned problems [3, 4].

28.3.2 Design M-RPS Based Customized Service

In a hospital, tags can be used for asset management for location finding. Patient tags are effective in preventing medical accidents, but must be properly designed and constructed to avoid massive collateral damage to user privacy. Hence, we define three-step privacy-aware service architecture for our mobile RFID-based medical application service [10]. The first step is setting the default level of access control over patient information in the default policy. The second step is user-controllable profile-based privacy protection, and the third step is auditable privacy management. Furthermore, we introduce a new RFID-based service model and mobile phone application [8, 9].

The mobile RFID reader requests for information related to a tag attached to a patient from the backend IS via the middleware system. The mechanism allows individuals to control who can access their personal information. For privacy management, we apply the proposed profile-based privacy management architecture by the addition of a privacy bit to the tag, which is a simple and cost effective mechanism. The privacy bit is the only reference for the privacy service. The

Table 28.1 Examples of a default privacy weight level

Privacy related people	Privacy weight	Privacy information
Doctor	L4–L9	Medical history treatment information
Nurse	L4–L9	Medical history treatment information
General doctor	L3–L7	Medical treatment
General nurse	L3–L7	Medical treatment
Family	L2–L6	Medical tracking information
Emergency agency	L2–L6	Medical tracking information
Others	LI	All cut off

medical RFID information server check the privacy guaranteed service or not from the privacy policy. To illustrate how the privacy policy works on the IS, let us consider its use in the application and content information system of the service provider. The privacy level is stored in M-RPS. The RFID code format for the application is defined in the mobile RFID application data format as standard. The default privacy level follows the privacy applied standard of each application service; and if there is no standard, the privacy level is determined based on the results of a privacy impact assessment. The privacy level consists of a 10-tuple of information, where ‘L = L1, L2, … L10’ as the default privacy policy. It also protected by a secure tag area and privacy server. We also define privacy weights for medical information, as shown in Table 28.1.

Classify the personal medical information by patient’s policy and make personal’s profile. The patient can control his privacy level. Encrypted information can be transferred between the hospital and an emergency transportation service in XML format with security (WS Security) and also can be subject to the standard access control technology for Web services (XACML) (Figs. 28.1 and 28.2).

In the proposed hospital data management system, RFID-tagged medical card are given to patients on registration. Patients with sensitive conditions, for example, heart disease or cerebral hemorrhage, can use the medical card to rapidly provide medical history that can used for fast application of first aid. Further, biosensors can be incorporated to provide real-time data to the doctor for each specific patient. The RFID patient tags also can be used to verify patient identity to ensure the correct treatment is administered. Thus, the system allows chartless service.

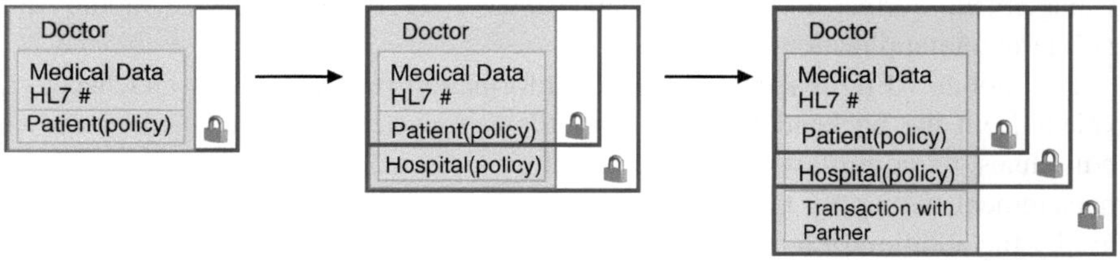

Fig. 28.1 Electronic signature and authentication

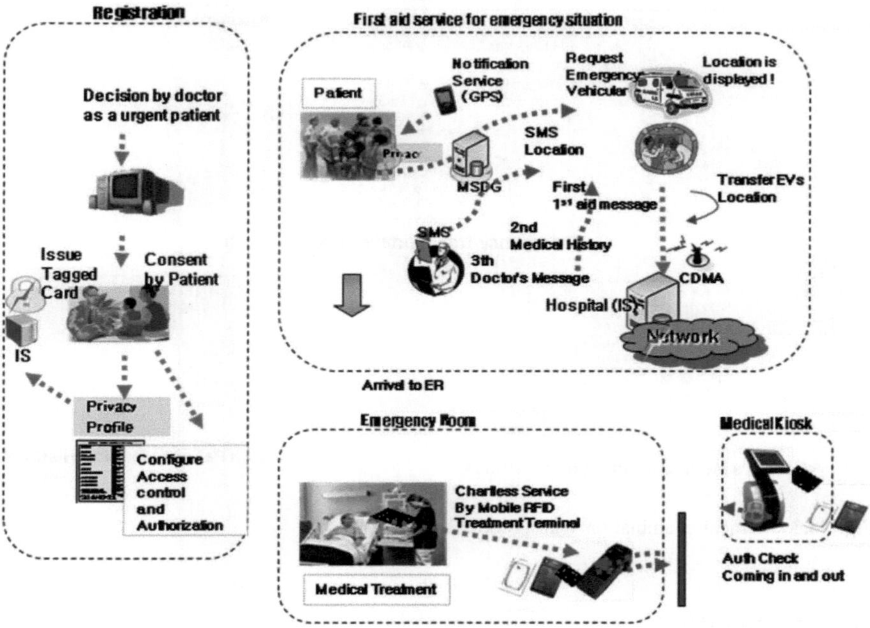

Fig. 28.2 Proposed customized ubiquitous hospital model

28.3.3 Implementation

The hospital generated an initial set of control data, which included the patient code, medical ID, and related information. The default privacy level was used and the patient was not allowed to control security policy. In order to provide authentication and privacy interface to patient as a agent in medical discovery gateway and hospital's information server system. Essentially, each bit of sensitive data was initially classified by the default privacy weight, which was then modified by the end user's detailed policy. The user-controllable privacy policy in this system evaluation is considered a basic part of RFID privacy management. The compatibility and scalability may be limited, which will hamper system migration, but the mechanism is suitable for policy based privacy control. The proposed privacy management mechanism was implemented in an actual medical emergency room, including a networked medical information RFID kiosk, RFID networked emergency rescue system, and medical examination service, as shown in Fig. 28.3. There is some approach applying the RFID to medicine and hospital. From above, proposed privacy scheme has advantages in custom centric approach aspect for constructing a privacy aware ubiquitous medical system.

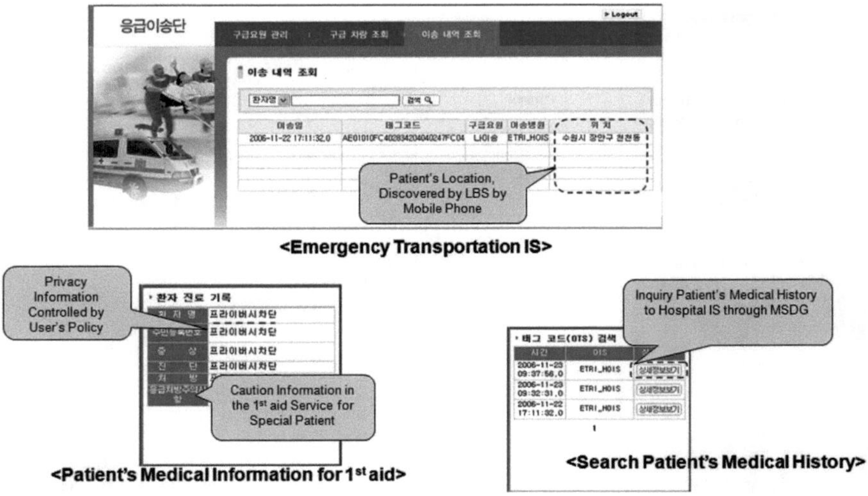

Fig. 28.3 Medical examination with proposed system

28.4 Conclusion

Mobile RFID readers are being actively researched and developed throughout the world, and more efforts are underway for the development of related service technologies. Though legal and institutional systems endeavor to protect privacy and encourage data protection, the science and engineering world must also provide suitable technologies. Seemingly, there are and will be no perfect security/privacy protection methods. The technologies proposed in this paper, however, would contribute to the development of secure and reliable RFID systems.

In this paper, we describe the security analysis and implementation leveraging globally networked mobile RFID-based wellness service in IoT Environment. We propose a secure mobile RFID-based wellness service framework leveraging mobile networking. The framework provides a means for safe use of mobile phone-based RFID services by providing security to personalized wellness service.

Acknowledgments This paper is extended and improved from accepted paper of UCSWSN 2014, ICHIT2011 Conferences. This research was supported by Basic Science Research Program through the National Research Foundation of Korea (NRF) funded by the Ministry of Education (2013R1A1A4A0101 3587).

References

1. Kang T, Lee H, Park D, Bang H, Park N (2013) Creation mechanism for access group based on user privacy policy-based protection. In: MUSIC, pp 125–130
2. Park N (2010) Security scheme for managing a large quantity of individual information in RFID environment. In: Zhu R, Zhang Y, Liu B et al (eds) Information computing and applications, vol 106. Springer, Berlin, pp 72–79
3. Park N (2011) Implementation of terminal middleware platform for mobile RFID computing. Int J Ad Hoc Ubiquitous Comput 8:205–219
4. Park N (2011) Secure UHF/HF dual-band RFID: strategic framework approaches and application solutions. In: ICCCI, vol 1, pp 488–496
5. Park N (2011) Customized healthcare infrastructure using privacy weight level based on smart device. In: ICHIT, vol 2, pp 467–474
6. Park N (2011) Secure data access control scheme using type-based re-encryption in cloud environment. In: Katarzyniak R, Chiu T, Hong C et al (eds) Semantic methods for knowledge management and communication, vol 381. Springer, Berlin, pp 319–327
7. Park N (2012) Cell phone-based mobile RFID: models, mechanisms and its security. Int J Radio Freq Identif Technol Appl 4:67–101
8. Park N (2012) Mobile RFID/NFC linkage based on UHF/HF dual band's integration in U-sensor network era. In: Park JH, Kim J, Zou D et al (eds) Information technology convergence, secure and trust computing, and data management, vol 180. Springer, Netherlands, pp 265–271
9. Park N, Kim M (2014) Implementation of load management application system using smart grid privacy policy in energy management service environment. Cluster Computing, vol 17, Springer, pp 653–664
10. Park N, Kwak J, Kim S, Won D, Kim H (2006) WIPI mobile platform with secure service for mobile RFID network environment. In: APWeb workshops, pp 741–748

References

Chapter 29
Interactive Drawing Based on Hand Gesture

Tae-Eun Kim

Abstract In this paper, a system that draws drawings on a real-time camera input image is proposed. First, a background subtraction operation is performed on the camera input image. A skin area detection algorithm is applied to the image that has undergone a background subtraction operation. After applying a labeling algorithm to the detected skin region image, a pen is augmented on the position of the hand. In addition, based on the obtained coordinates, a hand gesture-dependent drawing is drawn simultaneously on the camera image, in real-time.

Keywords Hand gesture · Interactive drawing · Augmented reality

29.1 Introduction

Attributable to the advances in computers, a variety of human computer interface (HCI) systems are being developed. This paper proposes a system that draws hand-drawn drawings on a real-time camera image via HCI. There are various drawing methods and there are many systems that provide assistance for creating drawings using the computer; however, such systems require devices such as a mouse or a tablet.

The proposed system facilitates easy drawing using only a computer and a camera. Furthermore, it provides augmentation in real-time to the drawing that the user is drawing right on the camera input image. In previous systems, the background is fixed when a drawing is drawn. The system proposed in this paper draws drawings right on the image inputted from the camera; therefore, it offers the user a greater sense that he/she is drawing in the same space that the user is occupying.

T.-E. Kim (✉)
Department of Multimedia, Namseoul University, Cheonan, Chungnam 330-707, South Korea
e-mail: tekim5@empas.com

J.J. (Jong Hyuk) Park et al. (eds.), *Future Information Technology - II*,
Lecture Notes in Electrical Engineering 329,
DOI 10.1007/978-94-017-9558-6_29

In order to enhance the sense of realism when drawing on a real-time screen in the proposed system, a pen is augmented on top of the hand. In previous studies, markers were used to augment virtual objects at specific locations [4]. However, in recent years, studies in which markers are not used in the augmented reality have been pursued. By using a specific color wristband so as to avoid using a specific marker, Choi et al. [2] conducted a study on hand tracking and augmentation. In this present paper, a pen is augmented without the use of a marker. Figure 29.1 shows the proposed system's schematic diagram. In this paper, a method in which the human hand is searched for and drawings are made on the screen in accordance the hand movement is proposed. Configuration of the system proposed in this paper is comprised of, in order: background subtraction, skin region detection, labeling, pen augmentation and figure synthesis.

29.2 Main Discussion

29.2.1 Background Subtraction

As images are inputted from the camera, first, background subtraction is performed. For background subtraction, a pre-treatment process transforms the camera input image and the reference image into a grey image. To accomplish background subtraction, first, the first frame of the camera input image is captured and background subtraction is applied, as this image becomes the reference image. There are two stages in the background subtraction operation. First, there is a step where the

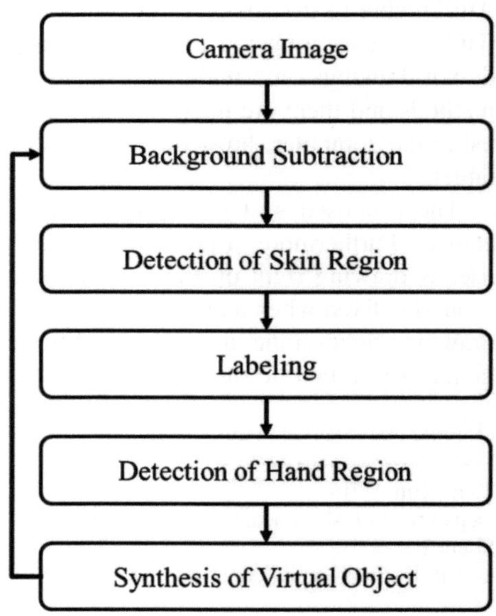

Fig. 29.1 Configuration (flowchart) of proposed system

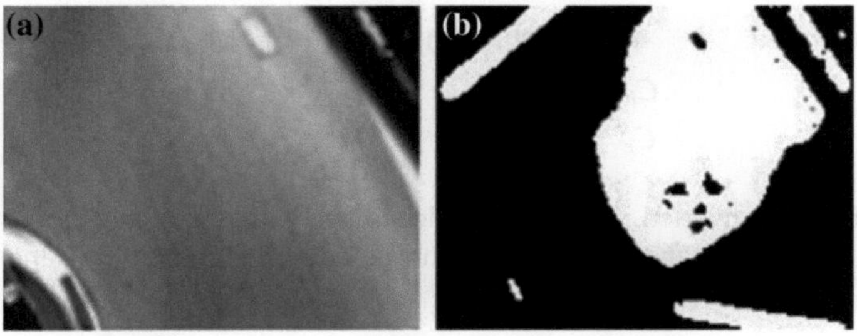

Fig. 29.2 **a** Reference image. **b** Difference image

camera input image is subtracted from the reference image. In addition, there is a step in which the reference image is subtracted from the camera input image [5]. The subtracted image from the two process steps is used to obtain the final difference image via OR logic operation. Equation (29.1) represents the subtraction operation. B1 and B2 represent the subtraction image of the two steps. The images of C and R represent the camera input image and reference image, respectively. B3 represents the operations of B1 and B2 (Fig. 29.2).

$$\begin{aligned} B_1(x,y) &= \begin{cases} 0, & if(C(x,y) - R(x,y) < threshold \\ 1, & otherwise \end{cases} \\ B_2(x,y) &= \begin{cases} 0, & if(R(x,y) - C(x,y) < threshold \\ 1, & otherwise \end{cases} \\ B_3(x,y) &= B_1(x,y) \cup B_2(x,y) \end{aligned} \tag{29.1}$$

29.2.2 Detection of Skin Region

To conduct detection of the skin region, a pretreatment process of background subtraction is carried out. In order to extract the hand from the camera input image that has undergone background subtraction, detection of the skin region is carried out. In performing extraction of the skin region, in the B3 image of Eq. (29.1), for parts whose pixel values that correspond to each pixel position are non-zero RGB color images inputted from the camera are shown. Detection of the skin region is carried out from the image inputted in such a way. For performing detection of the skin region, Eq. (29.2) is used to convert the RGB color image into YCbCr color image [6].

$$
\begin{aligned}
Y &= 0.299R + 0.587G + 0.114B \\
C_b &= B - Y \\
C_r &= R - Y
\end{aligned}
\tag{29.2}
$$

In the image converted into YCbCr color image with the values of Cb and Cr as reference, detection of the skin region is carried out. Equation (29.3) is the reference region of Cb and Cr [3].

$$
\begin{aligned}
77 &\le C_b \le 127 \\
133 &\le C_r \le 173
\end{aligned}
\tag{29.3}
$$

Figure 29.3 shows the extracted skin region using Eq. (29.3).

29.2.3 *Labeling*

Looking at the skin region extraction image in Fig. 29.3b, it can be seen that not just the person's skin but also other objects are detected as well.

This is due to the fact that a skin color detection algorithm was used in the skin region detection color image. As such, in order to eliminate other objects that have colors similar to the skin color, a labeling and post-treatment processes are performed.

In this paper, for real-time processing One-Pass run length labeling algorithm was used as a labeling algorithm. In order to carry out One-pass run length labeling, a conversion process for the final image that has undergone skin region detection is needed. In the image for which labeling was performed, the determined regions are sorted based on the size of each region and the largest size region is determined as the hand. One-Pass run length labeling algorithm has faster speed of operation than the existing Grass-fire labeling method. Whereas Grass-fire labeling method detects from the starting pixel of a region to the finishing pixel in one time, One-Pass run length labeling algorithm uses a method of detecting one line at a time in an image and, after one line is detected, while detecting the next line it assigns a sequential

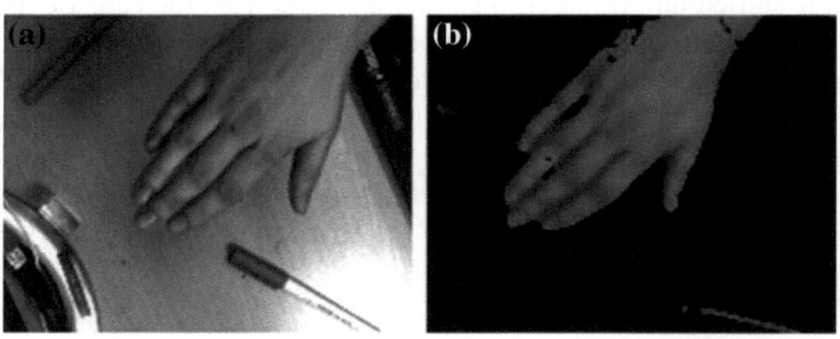

Fig. 29.3 **a** Real-time camera image. **b** Skin region extraction image

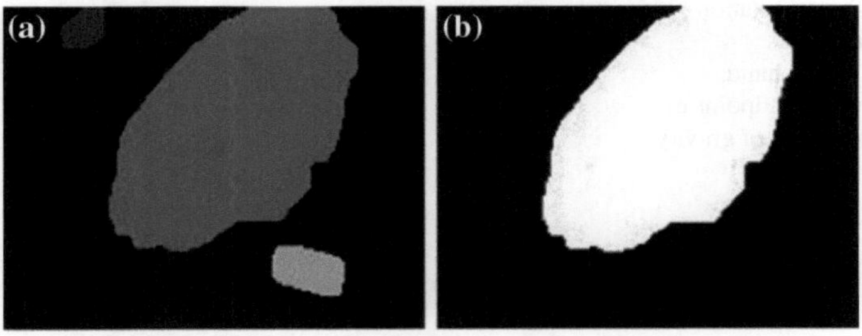

Fig. 29.4 **a** Labeling image. **b** Image after size comparison

number to the region in each line that is deemed as an object. For regions determined in such a way, by comparing with the determined regions of upper lines, when there are contiguous regions this algorithm makes a final determination that the regions are the same region [1]. Figure 29.4a is the image after labeling the skin region detection image, and Fig. 29.4b is the image that underwent only hand region extraction through size comparison after labeling.

29.2.4 Synthesis of Hand Tracking and Virtual Objects

With the detected hand image through size comparison after labeling as the reference, for the synthesis of drawings the center of gravity of the hand (c) and the endpoint of the hand (a) are first obtained. When the center of gravity of the hand (c) and the endpoint of the hand (a) are obtained, a point between the two points is detected by using Eq. (29.4).

$$\begin{aligned} Mp.x &= (CP.x - FP.x)/2 + FP.x \\ Mp.y &= (CP.y - FP.y)/2 + FP.y \end{aligned} \tag{29.4}$$

The point between the center of gravity (CP) and the endpoint of the hand (FP) is used as the reference coordinate for representing the drawing through the hand on the real-time camera input image and in addition it becomes the reference coordinate for pen augmentation. The reason for augmenting the pen with the point between the center of gravity (CP) and the endpoint of the hand (FP) and representing the drawing is due to the position of the pen tip when a person is holding the pen. Figure 29.5 shows an image representing all the points.

The superscript numeral used to refer to a footnote appears in the text either directly after the word to be discussed or—in relation to a phrase or a sentence—following the punctuation mark (comma, semicolon, or period). Footnotes should appear at the bottom of the normal text area, with a line of about 5 cm set immediately above them.

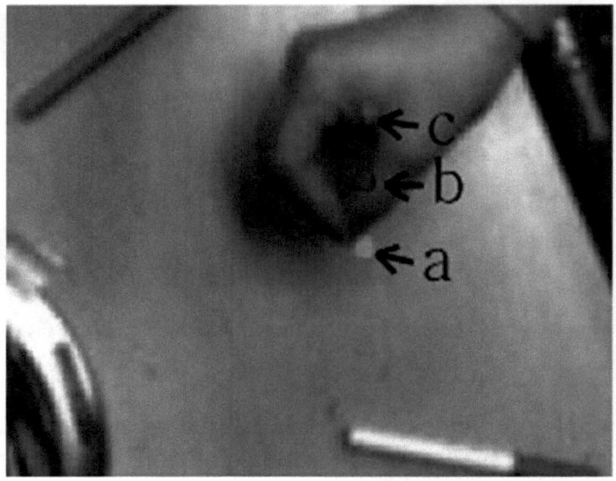

Fig. 29.5 Representation image of the points: *a* endpoint of the hand, *b* point between endpoint of the hand and center of gravity, *c* center of gravity

29.3 Experimental Results

One 6 mm lens 1394b camera was used in this study. The first frame acquired from the real-time camera image is set as the reference image. The reference image is used in background subtraction in two steps; which entails, first, subtraction of the camera input image from the reference image, and a method of subtracting the reference image from the camera input image in a different step, and by using OR logical operation, the background subtraction image is obtained. Detection of the skin region is carried out on the parts whose pixel values of the acquired background subtraction image are non-zero. In order to solve the issue of detecting as regions the colors that are similar to the skin color during detection of the skin region, after labeling the hand region is extracted through comparison of region sizes. From the extracted hand region the center of gravity and endpoint of the hand are detected, and afterwards the point between the two points is detected. With the point detected in this way as the reference the pen is augmented and a drawing that follows the hand movement is drawn on the real-time camera input image.

Figure 29.6 shows images of implementation of the system. In Fig. 29.6, it can be seen that a penis drawn over the pen, and the pen is augmented with the center of gravity and the endpoint of the hand as the reference. The highlighted portions (black circles) in Fig. 29.6a are the problem areas caused by a flickering phenomenon of the reference coordinate for making a drawing of the area indicated reference coordinates for drawing the picture blur due to duplicate the problem was being drawn. Figure 29.6b shows a human face drawn on the desktop through the system.

Fig. 29.6 **a** Augmentation of pen. **b** Proposed sketch system

29.4 Conclusion and Future Research Directions

In this paper, the hand was estimated from the camera input image and, by tracking the hand position, three coordinates of the hand were extracted [1]. With the extracted coordinate values as the reference, a drawing, which follows pen augmentation and hand gestures, was drawn. Although a slight flickering phenomenon was seen, the results of overall, stable augmented reality were obtained through an experiment. The slight flickering phenomenon seen in the existing system will be compensated for in future studies. In addition, a research that represents drawings drawn on the screen in a more realistic 3D will be pursued.

Acknowledgments Funding of this paper was provided by Namseoul University.

References

1. Anton H (1987) Elementary linear algebra. Wiley, Hoboken
2. Choi JY, Park HH, Park JI (2008) Interface for augmented reality using efficient hand gesture recognition (FCV), pp 439–444
3. Gasparini F, Schettini R (2006) Skin segmentation using multiple thresholding. In: Proceedings of SPIE internet imaging VII, vol 6061, pp 60610F.1–60610F.8
4. http://www.hitl.washington.edu/artoolkit/
5. Piccardi M (2004) Background subtraction techniques: a review. Int Conf Syst Man Cybern 4:3099–3104
6. Vezhnevets V, Sazonov V, Andreeva A (2003) A survey on pixel-based skin color detection techniques. In: Proceedings of Graphicon-2003, Moscow, Russia, pp 85–92

Chapter 30
On the Performance of Quasi-Orthogonal Space Time Block Coded Massive MIMO with up to 16 Antennas

Khin Zar Chi Winn, Phyu Phyu Han, Kasun Bandara and Yeon-Ho Chung

Abstract Massive multiple-input multiple-output (MIMO) using a large number of antennas at both transmitter and receiver sides based on quasi-orthogonal space time block code (QOSTBC) is presented. Space-time block code (STBC) is a MIMO transmit strategy that applies transmit diversity and high reliability. QOSTBC is attractive because it achieves higher code rate than orthogonal STBC and lower decoding complexity than non-orthogonal STBC. We present the performance of massive MIMO systems using the QOSTBC with multiple antennas up to the 16 × 16 configuration. The performances of 2 × 2, 4 × 4, 8 × 8 and 16 × 16 massive MIMO systems have been presented. Simulation results show that the massive MIMO systems with QOSTBC give significant performance improvement with full rate and full diversity, compared with previously considered massive MIMO systems.

Keywords Massive MIMO · Quasi-orthogonal STBC · Full rate

30.1 Introduction

The demand for mobile communication systems with high data rates has significantly increased in recent years. Many technical challenges remain in designing robust wireless networks that deliver the performance necessary to support

K.Z.C. Winn · P.P. Han · K. Bandara · Y.-H. Chung (✉)
Department of Information and Communications Engineering, Pukyong National University, Busan, Republic of Korea
e-mail: yhchung@pknu.ac.kr

K.Z.C. Winn
e-mail: khinzar111@gmail.com

P.P. Han
e-mail: phyu2han@gmail.com

K. Bandara
e-mail: kassae6@gmail.com

J.J. (Jong Hyuk) Park et al. (eds.), *Future Information Technology - II*,
Lecture Notes in Electrical Engineering 329,
DOI 10.1007/978-94-017-9558-6_30

emerging applications [4]. Multiple-input multiple-output (MIMO) systems offer numerous benefits over conventional single-input single-output (SISO) systems, such as the potential to facilitate data rates considerably higher or to significantly improve the reliability of a wireless link. In fact, MIMO technology is envisioned to be a core transmission technology for high speed wireless communications, due to increased data rate and performance obtainable largely from transmit diversity and spatial multiplexing [2].

MIMO scheme can be split into two categories: space time coding (STC) and spatial multiplexing (SM). STC improves the reliability while SM increases data rate [5]. STC, introduced by Tarokh and Calderbank [12], is promising method where the transmitted code symbols per time slot are equal to the number of transmit antennas. The class of linear STBC is the major category of STC and can be divided into subclasses, such as orthogonal STBC (OSTBC) [1] and quasi-orthogonal STBC (QOSTBC) [6] that is typically designed for more than two antenna systems with increased decoding complexity.

The massive MIMO is considered a candidate for a next generation wireless communication system [3]. For full benefits from the massive MIMO, the number of antennas is required to be at least more than 10 antennas at each side of the transmitter and receiver [8]. However, conventional performance analyses have been presented with only up to 8 antennas in the literature [8]. Therefore, a thorough report on a larger scale massive MIMO system still needs to be presented.

In this paper, we examine 2×2, 4×4, 8×8 and up to 16×16 massive MIMO systems. Comparative studies are undertaken for these different number of antenna configurations.

The rest of this paper is organized as follows. The details of massive MIMO system are described in Sect. 30.2. Section 30.3 describes STBC. In Sect. 30.4, the orthogonal designs (ODs) are explained. Section 30.5 describes quasi-orthogonal design. Section 30.6 present simulation results and conclusions are drawn in Sect. 30.7.

30.2 Massive MIMO System

A potential candidate for future mobile communication systems is massive MIMO, also known as Full-dimension MIMO or Hyper-MIMO [3]. This is a form of multi-user multiple antenna wireless communication which promises improvements in spectral efficiency and energy efficiency over 4G technology. A recognized feature of massive MIMO is that a large number of antennas can function for a significantly smaller number of active autonomous terminals. Large antenna arrays can potentially reduce uplink (UL) and downlink (DL) transmit power through coherent combining with increased antenna aperture [10]. Most recently, massive MIMO has been investigated as new cellular network architecture with several attractive characteristics [9].

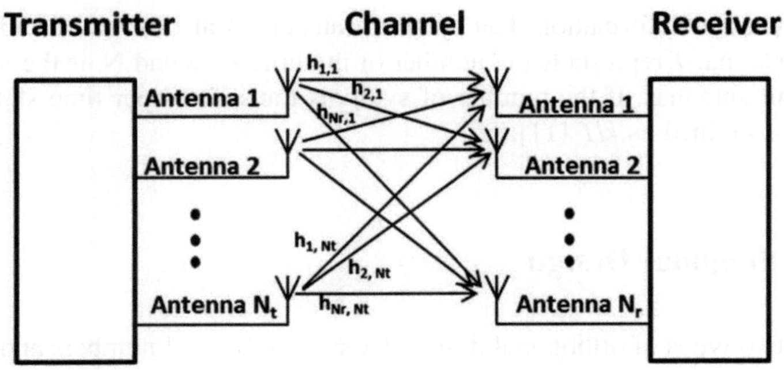

Fig. 30.1 Block diagram of massive MIMO system

Let N_t and N_r denote the number of transmit and receive antennas, respectively. The channel with N_r outputs and N_t inputs is denoted as a $N_r \times N_t$ matrix. A massive MIMO usually employs a large number of transmit and receive antennas, at least larger than 10 antennas. Figure 30.1 depicts a massive MIMO system block diagram. We consider a transmission scheme with multiple transmit and receive antennas. The MIMO system model can be represented as

$$x = Gz + w, \tag{30.1}$$

where $x \in C^{(N_r \times 1)}$ and $z \in C^{(N_t \times 1)}$ denote the received and transmitted complex vector, respectively. $G \in C^{(N_r \times N_t)}$ is the complex channel matrix and $w \in C^{(N_r \times 1)}$ is noise vector.

30.3 Space-Time Block Code

Space-time block codes (STBC) are designed to obtain the maximum diversity order for the provided number of transmit and receive antennas. These codes are orthogonal and can achieve full transmit diversity specified by the number of transmit antennas. The STBC matrix has columns equal to the number of the transmit antennas and rows equal to the number of the time slots required to transmit the data. The matrix representation of OSTBC, D, is expressed in (30.2).

$$D = \overset{\text{Transmit antennas} \longrightarrow}{\begin{bmatrix} a_{11} & a_{12} & \cdots & a_{1N_t} \\ a_{21} & a_{22} & \cdots & a_{2N_t} \\ \vdots & \vdots & \ddots & \vdots \\ a_{T1} & a_{T2} & \cdots & a_{TN_t} \end{bmatrix}} \Big\downarrow \text{Time slots} \tag{30.2}$$

where a_{ij} denotes information data symbol transmitted at time slot i and from jth transmit antenna. T represents the number of the time slots and N_t is the number of the transmit antennas. If the number of symbols transmitted per time slot is k, the code rate is defined as k/T [11].

30.4 Orthogonal Design

There are two types of orthogonal design theory: one for real numbers and other for complex numbers. For real or complex numbers, matrices satisfying the following equation are called orthogonal designs [7].

$$D^{H}D = \left(|a_1|^2 + |a_2|^2 + \cdots + |a_i|^2\right)I, \tag{30.3}$$

where D^{H} is the Hermitian of the matrix D, if D has complex numbers. D^{H} may also represent the transpose of the matrix D, if D has real numbers. In the real orthogonal designs, an $N_t \times N_t$ matrix is with real entries $\pm a_i$, $i = 1,\ldots, N_t$. In the complex orthogonal designs, an $N_t \times N_t$ matrix is with complex entries $\pm a_i, \pm a_i^*, \pm j a_i$, and $\pm j a_i^*$, $i = 1,\ldots, N_t$.

In complex orthogonal design, not only do the information symbols appear but also their conjugates. The ith column of the code matrix can contain either the information symbols $a_1,\ldots, a_T$, or their conjugates $a_1^*, \ldots, a_T^*$ only.

The most popular OSTBC is the Alamouti code [1]. This code uses a complex orthogonal design with full diversity at full rate. Alamouti code matrix can be shown as

$$\mathrm{D} = \begin{bmatrix} a_1 & a_2 \\ -a_2^* & a_1^* \end{bmatrix} \tag{30.4}$$

where * denotes complex conjugate. Here, two time slots are required to transmit two symbols. Thus, $k = 2$ and $T = 2$. Hence, the code rate of Alamouti Code is 1. However, when three or more transmit antennas are employed, the maximum transmission rate of the complex valued OSTBC with linear processing reduces to 3/4 [1].

r_1 and r_2 are the received signals at time t and $t + T$. Conjugating the signal r_2, the received signal can be expressed as (30.5)

$$\begin{aligned} r_1 &= h_1 a_1 + h_2 a_2 + \tilde{n}_1 \\ r_2^* &= -h_1^* a_2 + h_2^* a_1 + \tilde{n}_2. \end{aligned} \tag{30.5}$$

Therefore, the Eq. (30.5) can be described as

$$\begin{bmatrix} r_1 \\ r_2^* \end{bmatrix} = \begin{bmatrix} h_1 & h_2 \\ h_2^* & -h_1^* \end{bmatrix} \begin{bmatrix} a_1 \\ a_2 \end{bmatrix} + \begin{bmatrix} \tilde{n}_1 \\ \tilde{n}_2 \end{bmatrix} \tag{30.6}$$

or in short notation:

$$y = Ha + \tilde{n}, \tag{30.7}$$

where $y = \left[r_1, r_2^*\right]^{\mathrm{T}}$ represents the receive vector. For MIMO channel matrix, the rows and columns of the virtual channel matrix H of Alamouti STBC are orthogonal:

$$HH^{\mathrm{H}} = H^{\mathrm{H}}H = \left(|h_1|^2 + |h_2|^2\right)I_2 = |h|^2 I_2, \tag{30.8}$$

where I_2 is the (2×2) identity matrix and h^2 is the power gain. It is certain that the equivalent virtual channel matrix [7] depends on the structure of the code and the channel coefficients. The channel coefficients h_1 and h_2 can be used by the decoder as channel state information (CSI) if they can be absolutely estimated at the receiver.

30.5 Quasi-Orthogonal Design

Orthogonal space-time block code (OSTBC) can provide full transmit diversity with simple linear maximum-likelihood (ML) decoding complexity. In spite of these advantages, OSTBC has a code rate that is less than one when the number of transmit antennas more than two [12]. Therefore, the quasi-orthogonal space-time block codes are considered at the expense of slight orthogonality [7]. Each row and column contains all elements of a as shown in (30.9). Any element in a code word may occur with a positive or negative sign. The conjugate complex operation of symbols is only allowed on entire rows of the block matrix. This is required for holding quasi-orthogonality and low decoding complexity. Quasi-orthogonal STBC coding matrix is similar to the orthogonal space-time codes [13]. However, any column vector in encoding matrix of the quasi-orthogonal space-time block codes is related to only one of the columns. In this quasi-orthogonal code design, the columns of the transmission matrix are divided into groups. While the columns within each group are not orthogonal to each other, different groups are orthogonal to each other. The most famous quasi-orthogonal space-time block code is proposed by

Jafarkhani [6]. For 4 transmit antennas, a code with symbol transmission rate 1 was constructed from the Alamouti scheme as shown in (30.9):

$$D_4 = \begin{bmatrix} a_1 & a_2 & a_3 & a_4 \\ -a_2^* & a_1^* & -a_4^* & a_3^* \\ -a_3^* & -a_4^* & a_1^* & a_2^* \\ a_4 & -a_3 & -a_2 & a_1 \end{bmatrix} \tag{30.9}$$

OSTBC is decoded by decoding symbol one by one, and quasi-orthogonal STBC is decoded by decoding two pairs of symbols. The full transmission rate is useful and important for lower signal to noise ratios (SNRs) and higher bit error rates (BERs). Full diversity is efficient for higher SNRs and lower BERs. Since lower SNR range is of practical interest, QOSTBCs have attracted much attention recently.

30.6 Simulation Results

In order to evaluate a large massive MIMO, we have conducted performance evaluation of the QOSTBC based massive MIMO under additive white Gaussian noise (AWGN) channel and with ideal channel state information. We analyze the bit error rate (BER) performance of the massive MIMO system. It is assumed that all antennas in the system use the same energy level. The BER comparison has been undertaken according to the number of antennas. Note that it is possible to create different matrices that obey the quasi-orthogonal property. QOSTBC can achieve the full transmit diversity of $N_t = 2$, 4, 8 and 16 and code rate is 1.

The BERs of QOSTBC for 2 and 4 transmit and receive antennas are plotted in Fig. 30.2a and 8 and 16 transmit and receive antennas are plotted in Fig. 30.2b.

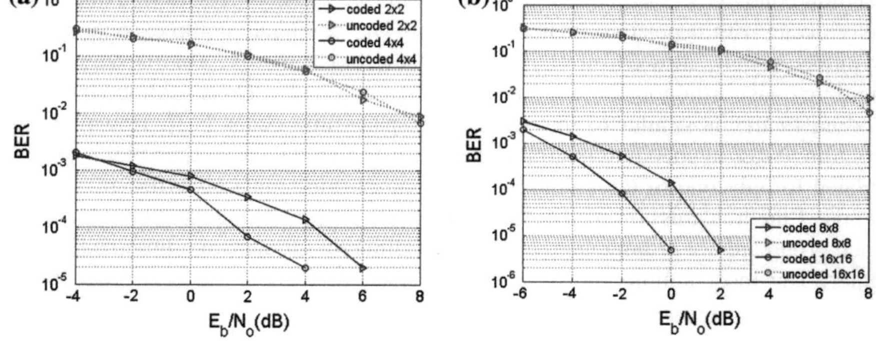

Fig. 30.2 Performances of massive MIMO systems: **a** 2 × 2 and 4 × 4 antennas, **b** 8 × 8 and 16 × 16 antennas

When the number of antennas increases, the BER performances of the system dramatically improve. For a large antenna configuration such as 16 × 16, the performance gain increases further. This is due to the fact that increasing antennas in massive MIMO provide significant increase in the BER performance through a diversity effect.

30.7 Conclusions

A large massive MIMO system with up to 16 × 16 antenna configuration has been evaluated. For a massive MIMO, performance analysis as well as energy efficiency must be conducted in a large antenna configuration. By making use of quasi-orthogonal space-time block code (QOSTBC) with full diversity and full rate, the simulation results show that the 16 × 16 configuration provides the best performance among other considered configurations. A more thorough analysis may need to be undertaken with realistic channel conditions.

References

1. Alamouti SM (1998) A simple transmit diversity technique for wireless communications. IEEE J Sel Areas Commun 16(8):1451–1458 (IEEE Press)
2. Bliss DW, Chan AM (2005) MIMO wireless communication. Lincoln 15(1):97–126
3. Cornell University Library. http://arxiv.org/abs/1304.6690
4. Foschini G (1996) Layered space-time architecture for wireless communication in a fading environment when using multi-element antennas. Bell Labs Tech J 1(2):41–59
5. Foschini GJ, Gans MJ (1998) On limits of wireless communications in a fading environment when using multiple antennas. Wireless Pers Commun 6:311–335 (Kluwer Academic, The Netherlands)
6. Jafarkhani H (2001) A Quasi-orthogonal space time block code. IEEE Trans Commun 49 (1):1–4 (IEEE Press)
7. Jafarkhani H (2005) Space-time codes: theory and practice. Cambridge University Press, Cambridge
8. Lee G, Park J, Sung Y, Seo J (2012) A new approach to beamformer design for massive MIMO systems based on k-regularity. In: IEEE Globecom workshops. IEEE Press, Anaheim, pp 686–690
9. Marzetta TL (2010) Noncooperative cellular wireless with unlimited numbers of base station antennas. IEEE Trans Wirel Commun 9(11):3590–3600 (IEEE Press)
10. Ngo HQ, Larsson EG, Marzetta TL (2013) Energy and spectral efficiency of very large multiuser MIMO systems. IEEE Trans Commun 61(4):1436–1449 (IEEE Press)
11. Su W, Xia XG (2004) A systematic design of high-rate complex orthogonal space-time block codes. IEEE Commun Lett 8:380–382 (IEEE Press)
12. Tarokh V, Calderbank AR (1999) Space-time block codes from orthogonal designs. IEEE Trans Inf Theory 45(5):1456–1467 (IEEE Press)
13. Weinrichter J, Rupp M (2005) Space time block coding for multiple antenna systems. http://publik.tuwien.ac.at/files/pub-et_10819.pdf

Chapter 31
Analysis of Gesture Combos for Social Activity on Smartphone

Meng-Yen Hsieh, Tien-Chi Huang, Jason C. Hung and Kuan-Ching Li

Abstract Due to sensing technologies, gestures are assisting users in conveniently operating smartphone. This paper offers the influence of gesture control with hand-held motions, since accelerometer sensor is easy for application and measurement. Gesture combos are designed to trigger a series of tasks, while a single gesture has been designed to execute a task on current APPs. We give an analysis of complex composite gestures, and explore user basic and social tasks on smartphone in order to find a match between social tasks and gesture combos. A series of tasks are illustrated, while some matching mechanisms are used for developing control functions with gesture combos. Some experiences of measuring gestures are also shown in the paper.

Keywords Gesture · Sensor · Smartphone

31.1 Introduction

Various sensor components are embedded on smartphone, providing new approaches to control mobile APPs, and broadening APP's domain cooperating with other electric devices. Smartphone could include a number of sensors based on hardware

M.-Y. Hsieh (✉) · K.-C. Li
Department of Computer Science and Information Engineering, Providence University, Taiwan, Taiwan
e-mail: mengyen@pu.edu.tw

K.-C. Li
e-mail: kuancli@pu.edu.tw

T.-C. Huang
Department of Information Management, National Taichung University of Science and Technology, Taiwan, Taiwan
e-mail: tchuang@nutc.edu.tw

J.C. Hung
Department of Information Technology, Overseas Chinese University, Taiwan, Taiwan
e-mail: jhungc.hung@gmail.com

J.J. (Jong Hyuk) Park et al. (eds.), *Future Information Technology - II*,
Lecture Notes in Electrical Engineering 329,
DOI 10.1007/978-94-017-9558-6_31

design, such as accelerometer, magnetometer, light, proximity, gyroscope, barometer, temperature, and humidity sensors. Parts of the sensors have been used to develop APPs with gesture control. Some famous Android APPs have been released on Google Play market such as Air Call Accept and Hovering Controls to support gesture control for basic operations on smartphone. However, no APP applies gesture controls to social activities at present. This paper describe the analysis and experiences of gesture combos, while social tasks are comprised in APP development and possible matching mechanisms are presented for these gestures. To reduce the complexity in research about gesture combos, the only focus is on handheld-motion gesture controls for analysis and discussion.

Some APPs with different types of gesture controls are described as follows. Magus is a free 3D gesture launcher as one of hand-held control APPs in support of not only tilting, flipping, and running single gestures, but also personal combos combining with the single gestures. Gesture combos are arranged to trigger smartphone actions chosen by users. Hovering Controls as one APP of touchless gesture controls allows users performing particular operations through hovering hold, swiping once, or swiping twice without touching the screen. These operations could be as starting a APP, silencing alarm, controlling photo and video playbacks. Google Gesture Search is a touch-based APP and provides search functions on smartphone. Users have to draw down a symbol by touching the screen with fingers so that relative applications, setting, or contact lists can be shown up on the screen immediately.

After a general instruction about gesture-controlled applications and the objectives in this section, possible gestures for mobile behaviors and related works are presented in Sect. 31.2. Section 31.3 describes the challenges when developers design gesture combos for mobile APPs. The possible sequences composed of a gesture combo are detailed and analyzed, while people's favor is as a factor to include gesture combos. In Sect. 31.4, not only the experience of gesture measurement is shown, but also matching problems between social tasks and gesture controls are proposed for searching more suitable developments with gesture combos.

31.2 Related Works

In our previous research [3], single gesture controls are used for performing social tasks in P2P. After pairing with NFC between any two closed peers, one peer sends a photo to the other by performing a hand-held gesture, and the other receives the photo by performing another gesture. The performance of gesture controls instead of button pressure triggers the actions of data sharing. In addition, a social framework for mobile APPs is proposed with possible social sharing contests.

Ruiz et al. [5], have finished a study about a mapping between user motion gestures and some smartphone tasks. They explored a set of user-specified gesture, and explain the qualitative data and the taxonomy of user operations on smartphone.

Normal tasks into two categories corresponding to motion gestures were measures and analyzed. Three different gesture models with seven hand gestures were proposed by Xu et al. [6] to detect MEMS 3-axes accelerometers and to interpret sensing data in computer through Bluetooth connection. A recognition system is developed based on sign sequence and template matching methods, excluding the time consuming user-training process. The segmentation algorithm of the system generates the gesture codes form sensing data in order to compare with the standard patterns decided in advance. A gesture recognition framework [4] was able to interact seamlessly with desktop applications, while gathering gesture information from accelerometer sensor in smartphone. User can generate gestures to control remote desktop applications instead of general conventional input/output devices such as mouse, keyboard, and joystick. Bluetooth is adopted for data transmission between smartphone and desktop. As soon as raw data from accelerometer sensor is received, appropriate events with corresponding gesture information will be filtered and processed.

31.3 Design Challenges

The two problems are significant for designing combos. First, what combos are suitable to users? And what social tasks could be required to complete social activities?

31.3.1 Gesture Combo

Developers consider some main problems for usage, while gestures combos are applied. First, users cannot memorize all possible gestures or their combos setting on smartphone. Users have different preferences for gesture combos. According to the result of psychological researches, people only learn few gestures and keep them in the mind. They remember easy and single gestures to control smartphone, better than complex and composite gestures. One gesture combo is consisting of a number single gestures, probability different to be memorized by users. Besides, to finish a gesture combo during user performance should be more difficult than to perform a single gesture. Consequently, to ask users for remembering many kind of complex gestures or unmemorable gesture composites will be unable on APP development.

A number of gesture combos are generated, even if APPs only supports few single gestures for controlling smartphone operations. We assume that a set of single gestures, defined as {Gestures} is designed for hand-held gesture control in one mobile APP. The number of single gestures is defined as |{Gestures}| and one gesture combo is organized from a fixed amount of single gestures, defined as num. The number of gesture combos are denoted as follows:

$$(C_1^{|\{Gestures\}|})^{num}, \quad where\, num \geq 1,\ \{Gestures\} \neq \Phi \tag{31.1}$$

Suppose that one gesture combo is organized with a number of unrepeatable single, the number of all of possible gesture combos, defined as the following expression:

$$\prod_{i=0}^{num-1} (C_1^{|\{Gestures\}|-i}), \quad where\, 1 \leq num \leq |\{Gestures\}|,\ \{Gestures\} \neq \Phi \tag{31.2}$$

Although a lot of gesture combos are generated, only parts of them could be designed in APPs due to user preferences. Besides the problem of user memory ability, developer must probe the coherence problem of that users can be easy to perform a combo. To determine whether a gesture combo is available for development, user preference for each single gesture and user acceptance for each combo with any two single gestures are investigated according to application features.

31.3.2 Social Activity

Some of smartphone tasks for gesture controls have been divided into the two categories, actions and navigation-based tasks [5]. Each of them includes two subcategories to separate these tasks operating into the system and application on smartphone. To design the gesture control is easy if one gesture is recognized to match only one task relative to the basic operations. Oppositely, some difficulties are rising to match a gesture combo for social sharing activities, while serial social tasks is organized for gesture combos.

Currently, APP markets have announced a large number of social APPs in order to achieve various social activities f or mobile users. Most of those APPs are divided into the two categories. One type is social network APPs and the other is peer-to-peer sharing ones [1–3]. One or more gesture combos assist users in completing social activities, such as message posting, social pairing, data selection, etc. Table 31.1 illustrates a number of basic tasks with user social sharing. To achieve a social sharing activity has to finish one or more of these tasks. For example, users want to post a file or message to social networks, denoted as a social sharing activity. Hence, developers design that gesture combos assist users to automatically and fast start the APP, login a social network, select a file, and post the file through some of the tasks such as Start, Login, Selection, and Post. In peer-to-peer sharing interaction, data sharing between two close users is achieved possible with some of the tasks such as Start, Search, Pair, Send, and Receive.

Moreover, developers have to decide which two or more tasks are able to be consecutive for gesture control based on their development features.

Table 31.1 Illustration of tasks for social activities

Social tasks	Descriptions
Start	Start APP or a social session
Login	Login with authentication
Post	Post data on internet
Wireless	Make wireless connection
Selection	Select data
Search	Search peers
Pair	Pairing between the two peers
Send	Send data to a user
Receive	Receive data from a user
Page	Page operations (page enter, next page, previous page, first page, last page, etc.)

31.4 Gesture and Combo Matching Analysis

31.4.1 Gesture Analysis

To understand the sensing curve of single gestures can evaluate the difficulty of gesture combos. With ten gestures, the number of possible gesture combos only combining with two single gestures with repetition is equal to 100 according to formula (31.1). The number of the gesture combos without repetitive single gestures is equal to 90 according to formula (31.2). Whatever combination or permutation, the number of the gesture combos are a lot for development. Questionnaire is required to understand what kind of gestures is suitable and comfortable for usage. The questionnaire for each gesture is proposed, including two important questions: (1) the gesture is easy to perform, and (2) I can perform the gesture exactly. About 100 users are involved to answer the questionnaires. Each question has three levels between for full approval and complete rejection. Table 31.2 depicts not only the axle states of each of the ten gestures must be evaluated for gesture recognition, but also the Like percentages of users from the questionnaire. According to the analysis, Gesture Swing and Left-Right, are simple to be estimated with the variation of only one axle's curve. However, they are the main gestures that users like better. The others are estimated with the x-axle and y-axle curves.

Circle motion is one of single gestures that users favored. However, its recognition is difficult more than others with only one axle recognition. Figure 31.1 describes three curve lines stating statistics from sensing data of the accelerometer. The curve lines of x- and y-axles have three and one decisive turning points, individually.

Table 31.2 Difficulty and like of single gestures

Single gesture	x-axle	y-axle	z-axle	Like (%)
Circle (clockwise, anticlockwise)	X	X		21
Cross	X	X		7
Triangle	X	X		3
Rectangle	X	X		2
SemiCircle (down, or up)	X	X		15
Swing-left/right	X			22
Swing-down/up			X	19
Hand rotation (180°)			X	6
Left-right	X			5
Infinite	X	X		0

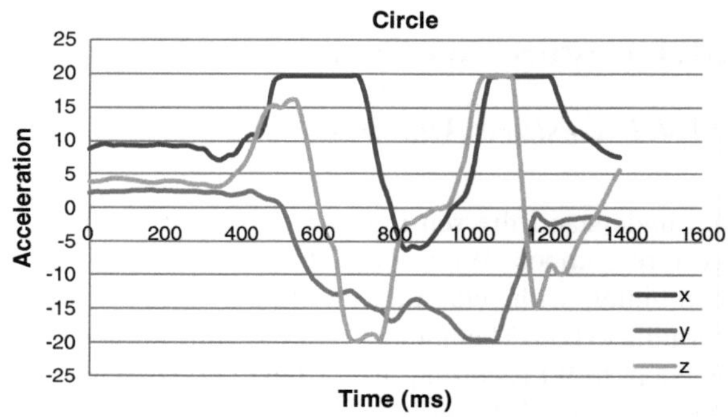

Fig. 31.1 *Curve lines* from a circle gesture

31.4.2 Combo Matching

When carrying out a number of consecutive tasks, developers must match gesture combos that users favored most for the task combos happened frequently. Based on Table 31.1, Fig. 31.2 illustrates a relationship network where nodes are tasks with direction edges denoting their sequences for P2P social activities based where one user can send (or receive) data to (or from) another. Degree between two tasks is used to describe the relations. Degree calculation is a good measure of the total connections a task has. The higher the degree, the more important the task is. For the reason, it is a good matching approach where the gesture combos that users prefer are corresponding to the task combos starting with the task of the high degree. The in-degree and out-degree of Node's Page are 6 and 2 individually. The in-degree and out-degree of Node's Search are the same degree, 3.

Search and Pair have the same out-degrees higher than the others. The possible pairs of consecutive tasks are (Search, Search), (Search, Page), (Search, Pair), (Pair, Pair), (Pair, Page), and (Pair, Receive). Circle, Swing, and SemiCircle are the three favorite gestures in the questionnaire. Suppose that Gesture Circle, Swing,

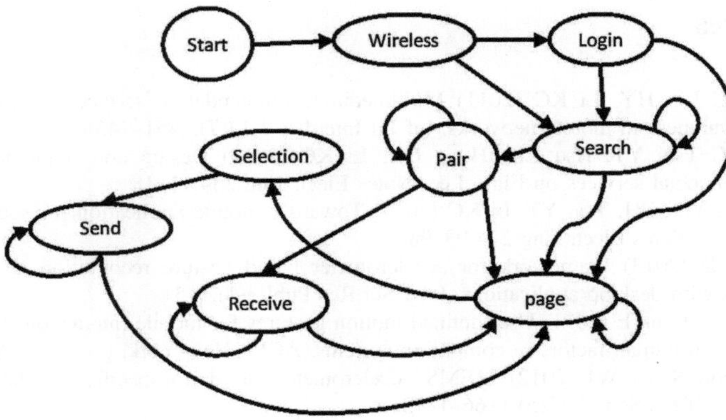

Fig. 31.2 Illustration of an order of social tasks

SemiCircle are matching to Task Search, Page, Pair, individually. Since a swing gesture always comes up with other actions such as left, right, up, and down, Page with a set of sub-tasks, for example, previous page or next page, is suitable to Swing more than the others. Possible matching states are proposed as follows.

1. Swing-Left, Swing-Right, Swing-Up, Swing-Down are corresponding to Next-page, Previous-Page, Send, Receive.
2. (Circle, Circle), (Circle, Swing), and (Circle, SemiCircle) are corresponding to (Search, Search), (Search, Page), and (Search, Pair).
3. (SemiCircle, SemiCircle), (SemiCircle, Swing), (SemiCircle, Swing-Down) are corresponding to (Pair, Pair), (Pair, Page), and (Pair, Receive).

31.5 Conclusion

This research discusses gesture combos on triggering social activities, while handheld gesture controls are applied to smartphone. Since the combinations of single gestures is numerous and difficult to be memorized by users, only gesture combos consisting of two gestures are presented. The design challenge is proposed for matching gesture combos to social tasks. Some social tasks from social APPs are illustrated with their relations, while the sequence of operations during gesture controlling is involved. Although the matching mechanism is not only one solution, the experience provides references to the development of gesture combos for mobile social activities.

Acknowledgments This investigation was supported by Ministry of Science and Technology, Taiwan, under grant No. NSC102-2221-E-126-003.

References

1. Hsieh MY, Lin HY, Li KC (2011) Multimedia recommendation services based on social context awareness in mobile networks. Inf Int Interdisc J 14(7):2451–2458
2. Hsieh MY, Tsai YT, Hsu CH, Hung CH, Li KC (2013) Design and implementation of multimedia social services on Elgg. Lect Notes Electr Eng 234:11–18
3. Hsieh MY, Yeh CH, Tsai YT, Li KC (2014) Toward a mobile application for social sharing context. Lect Notes Electr Eng 274:93–98
4. Rajanna VD (2013) Framework for accelerometer based gesture recognition and seamless integration with desktop applications. Int J Sci Res Publ 3(1):1–5
5. Ruiz J, Li Y, Lank E (2011) User-defined motion gestures for mobile interaction. In: SIGCHI conference on human factors in computing systems. ACM, New York, pp 197–200
6. Xu R, Zhou S, Li WJ (2012) MEMS accelerometer based nonspecific-user hand gesture recognition. IEEE Sens J 12(5):1166–1173

Chapter 32
SOM Clustering Method Using User's Features to Classify Profitable Customer for Recommender Service in u-Commerce

Young Sung Cho, Song Chul Moon and Keun Ho Ryu

Abstract This paper proposes a SOM clustering method using user's features to classify profitable customer for recommender service in e-Commerce. In this paper, it is necessary for us to classify profitable customer with RFM (Recency, Frequency, and Monetary) score, to use the purchase data to join the customers using SOM with input vectors of different features, RFM factors in order to do recommender service in u-commerce, to reduce customers' search effort for finding items, and to improve the rate of accuracy. To verify improved performance of proposing system, we make experiments with dataset collected in a cosmetic internet shopping mall.

Keywords RFM · CF (Collaborative filtering) · SOM (Self-Organizing Map)

32.1 Introduction

Due to the advent of ubiquitous and IoT environment, it is becoming a part of our common life style that the demands for enjoying the wireless internet using intelligent portable device such as smart phone, tablet PC, and PDA are increasing anytime or anyplace without any restriction of time and place. Data mining is useful in finding knowledge from huge amounts of data. Deboeck and Kohonen describe how SOM (Self-Organizing Map) can be used for effective clustering and segmentation of

Y.S. Cho · K.H. Ryu (✉)
Database and Bioinformatics Laboratory, Computer Science in College of Electrical and Computer Engineering, Chungbuk National University, Cheongju, Korea
e-mail: khryu@dblab.chungbuk.ac.kr

Y.S. Cho
e-mail: youngscho@empal.com

S.C. Moon (✉)
Department of Computer Science, Namseoul University, Cheonan, Korea
e-mail: moon@nsu.ac.kr

J.J. (Jong Hyuk) Park et al. (eds.), *Future Information Technology - II*,
Lecture Notes in Electrical Engineering 329,
DOI 10.1007/978-94-017-9558-6_32

financial data [5]. Clustering algorithm is a kind of customer segmentation methods commonly used in data mining. In this paper, SOM network is applied to segment the purchase data to join user data and finally formed clusters by using the purchase data to join user data with different features, RFM factors in order to do recommender service in u-commerce. The recommender system helps customers to find easily items and helps the e-commerce companies to set easily their target customer by automated recommending process. Therefore, customers and companies can take some benefit from recommender system. The possession of intelligent recommender system is becoming the company's business strategy. A recommender system using RFM segmentation analysis technique to meet the needs of customers, it has been actually processed the research [1–4]. We can make the solution for an efficient purchase pattern clustering based on SOM. Finally, we can improve the performance of recommender system through SOM learning method based on the purchase data to show customer's buying patterns. The next chapter briefly reviews the literature related to studies. The Chap. 3 is described a new method for recommender system in detail, such as system architecture with sub modules, the procedure of processing the recommender, the algorithm for proposing system. The Chap. 4 describes the evaluation of this system in order to prove the criteria of logicality and efficiency through the implementation and the experiment. In Chap. 5, finally it is described the conclusion of paper and further research direction.

32.2 Relative Works

32.2.1 RFM

RFM is generally used in database marketing and direct marketing and has received particular attention in retail. RFM consists of three initial characters. R means recency—"How recently a customer has purchased?". F means frequency—"How often she purchases?". M means monetary—"How much does she spend?". The general way to use RFM model in customer behavior analysis is to sort the customer data by each dimension of RFM variables and then divide the data into five equal quintiles. For recency, the customer database is sorted by purchase dates by descending order. So, the top segment is given a value of 5 and the others are discerningly assigned of 4, 3, 2, and 1. For frequency and monetary, sorting customer visiting frequency data and the customer data related to the amount of the money spent in descending order, respectively. These three variables belong to behavioral variables and can be acted as the segmenting variables by observing customers' attitudes toward the product, brand, benefit, or even loyalty from the database. We can suggest that using average purchase amount instead of total accumulated purchase amount is better in order to reduce co-linearity of frequency and monetary. Finally, all customers are presented by 555, 554, 553, …, 111, which thus creates 125 (5 × 5 × 5) RFM cells. Moreover, the best customer segment is 555, while the worst customer segment is 111. Based on the assigned RFM behavior scores, customers can be classified into segments and their profitability can be further analyzed. The RFM score can be a basis factor how to determine purchasing behavior on the internet shopping mall, is helpful to buy the item which they really want by the personalized recommender [2, 8].

32.2.2 Neural Network

The SOM introduced by Kohonen [7], is an unsupervised learning algorithm for clustering. Also SOM is called as a neural networks model based on competitive learning. SOM can convert a high dimensional input space into a simpler low dimensional discrete map. It has two layers which are input and feature layers. We can cluster all elements by feature map with two dimensions. Firstly SOM performs clustering with input vector X and weight matrix W. The data point X_i is treated one at a time. Also the closest W_j to X_i is found by Euclidean distance, and then W_j is updated as the following [6].

$$W_k = W_j + \alpha(X_i - W_j) \tag{32.1}$$

where W_j and W_k are current and new weights. So W_k moves to X_i. This learning is repeated until given conditions such as change rate of weights and the number of repeat. In this paper, we can use the SOM learning algorithm [6], where M_j and M_k are current and new weights. So M_k moves to X_i. This learning is repeated until given conditions such as change rate of weights and the number of repeat. In this paper, we can use the SOM learning algorithm [6] in Table 32.1.

Table 32.1 SOM learning algorithm

```
Input   Set of N dimension vector, X // input node
Output  Subset of input data (M subsets)
begin
  Randomly initialize Mi = Mi1, Mi2 , ...Min for each node ;
   for (t=0; unless a stopping condition is reached; Increase t)
     for (for all input data)
       for (i=0 to M)
     Compute Di = || Xt − Mi(t) || ;
       endfor
     Find the winner j=i such that
            Di(t) is minimum for over all I ;
     Update the winner j (and its neighbors) ;
     endfor
endfor
end
```

32.3 Our Proposal for Recommender Service in u-Commerce

32.3.1 Clustering Method Using SOM to Classify Profitable Customer

This proposal SOM clustering method in this paper is better than k-means clustering the data directly, is depicted. First, a large result sets of prototyping for clustering user data (much larger than the expected number of output count, purchase pattern in clusters) is formed using the SOM or some vector quantization algorithm. We can apply a SOM clustering to purchase data to join user data in order to classify profitable customer with RFM score having RFM factors for recommender service in u-commerce. Finally the prototyping application is made and the prototyping result is classified to make clusters in order to classify profitable customer with RFM score. The system can use the code of classification (54 bits), demographic variables such as age, gender, an occupation, skin type, region and customer's RFM factors as input vectors for pre-processing so as to be possible to recommend the items with efficiency. The system can make clusters with neighborhood customer-group using a new clustering method, which is classified by the code of classification and customer's RFM score in customer information. The system can take the preprocessing task which is able to use the whole purchase data by each rank of the RFM score and then makes cluster of purchase data sorted by item category, joined cluster of user data called by customer DB, neighborhood user group [2]. As a matter of course, the system can use the whole purchase data (sale). After that, the system using SOM algorithm, can recommend the items by each rank of the RFM score. The SOM learning algorithm for clustering of user's information to join user's score is depicted in Table 32.2.

32.3.2 The Procedural Algorithm for Recommender Service

The system can search cluster selected by using the code of classification and customer's RFM score in users' information. It can scan the preference of brand item in cluster, suggest the brand item with the highest score in item category selected by the highest probability for preference of item category as the average of brand item. This system can create the list of recommendation with TOP-N of brand item with the highest score to recommend the item with purchasability efficiently. This system can recommend the items with efficiency, are used to generate the recommendable item according to the basic of loyalty of RFM factors through clustering method using SOM algorithm. It can recommend the associated item to TOP-N of recommending list. This system takes the cross comparison with purchase data in order to avoid the duplicated recommendation which it has ever taken.

Table 32.2 SOM learning algorithm for clustering of user's information to join user's score

```
Step 1  Initialize parameters of SOM model
// Representative pattern of bits
for demographic variable(54bits), RFM(15bits) factors)
Step 2  Set input value vector
Step 3  Calculate Output value
```

$$O_j = \sum_{i=0}^{n} w_{ij} x_i \quad (2)$$

```
Step 4  Select winner node
```

$$O_j^* = Max(O_j) \quad (3)$$

```
Step 5  Readjust connection weights
```

$$w_{ij}(t+1) = w_{ij}(t) + \alpha\,(x_i(t) - w_{ij}(t)) \quad (4)$$

```
Step 6  Completion of learning
// IF Reach the learning  cycles
then Make the result of SOM
otherwise GO to Step 2
Step 7  Calculate output value
Step 8  Calculate winner node
Step 9  Result of pattern
```

32.4 The Environment of Implementation and Experiment and Evaluation

32.4.1 Experimental Data for Evaluation

We made the implementation for prototyping of e-shopping mall which handled the cosmetics professionally and did experiments. It was the environment of implementation and experiments in Apache 2.2.14, j2sdk 1.7.0_11 as Java environment, JSP/PHP 5.2.12 as server-side script, JQuery*mobile, XML/XHTML4.0/HTML5.0/CSS3/JAVASCRIPT as client-side script, C#.net framework 2.0, jakarta-tomcat, apache 5.0.28 as web server under Windows O.S.

32.4.2 Experimental Data for Evaluation

We used 319 users who have had the experience to buy items in e-shopping mall, 580 cosmetic items used in current industry, 1,600 results of purchase data recommended in order to evaluate the proposing system [3]. In order to do that, we make the implementation for prototyping of the internet shopping mall which handles the cosmetics professionally and do the experiment. We have finished the system implementation about prototyping recommendation system. We'd try to carry out the experiments in the same condition with dataset collected in a cosmetic internet shopping mall. It could be evaluated in MAE and output count by RFM score level for the recommendation system in clusters. It could be proved by the experiment through the experiment with learning data set for 12 months, testing data set for 3 months in a cosmetic cyber shopping mall [3]. The 1st system of SOM clustering method using user's features based on purchase data to join user data in order to classify profitable customer with RFM score having RFM factors, is proposing system called by "SOM", the 2nd system is other previous system (k-means) using k-means clustering algorithm based on the whole data.

32.4.3 SOM Results for Application of Neural Network

SOM network is applied to classify purchase data to join the user data, finally forms clusters of purchase pattern groups of user data with different features, demographic variables and RFM factors as input vectors. In order to segment purchase data join to user data into appropriate number of clusters, SOM is applied to determine the number of clusters. Figures 32.1 and 32.2, nine clusters are recommended among 1,600 purchase data, 319 customers when recency, frequency, monetary are the three input variables and then divide the data into five equal quintiles beside off

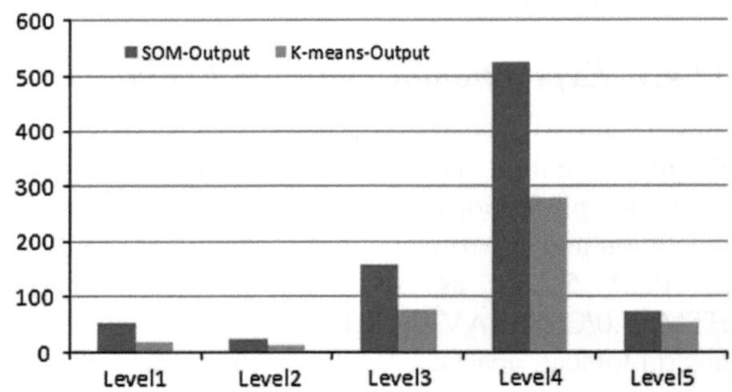

Fig. 32.1 Descriptive statistics of output counts by the level of RFM score with comparing SOM and k-means

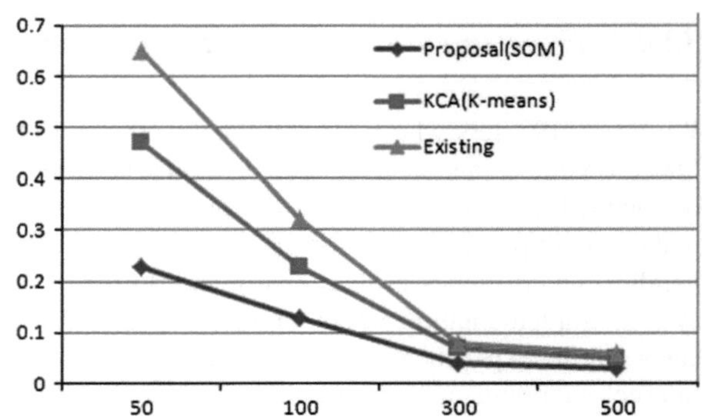

Fig. 32.2 The result of MAE by comparing proposal system (SOM) with previous system (k-means)

demographic variables. From the SOM result, we can find 5 level based on RFM score of customer so as to recommend the items in real-time environment [4]. The following figures show the result with statistics of output counts based on purchase data for the segmentation as comparing SOM and k-means. It is depicted in the result, that level 1 is the RFM score of customer is more 90 points, level 2 is the range of RFM score (score ≥ 80 and score < 90), level 3 is the range of RFM score (score ≥ 60 and score < 80), level 4 is the range of RFM score (score ≥ 40 and score < 60), and level 5 is the range of RFM score (score ≥ 20 and score < 40). The purchase data is not at the range of RFM score (score < 20). It shows the improvement on the number of output purchase pattern count in the result of evaluation levels for the proposal system (SOM) comparing with previous system called by "KCA" (k-means). The proposal is higher on the number of output purchase pattern count than the previous system from level 1 to level 5. As a result of that, the performance of the proposal system is improved better on the number of output purchase pattern count than previous system from level 1 to level 5.

32.4.4 Experiment and Evaluation

The proposing system's overall performance evaluation was performed by dividing the two directions. The first measurement is output counts of purchase pattern in the Table 32.2. The second evaluation is mean absolute error (MAE). The MAE between the predicted ratings and the actual ratings of users within the test set. The MAE is computed the following expression (32.5) over all data sets generated on purchased data.

$$\mathrm{MAE} = \frac{\sum_{i=1}^{N} |\epsilon_i|}{N} \tag{32.5}$$

N represents the total number of predictions, ε represents the error of the forecast and actual phase i represents each prediction (Fig. 32.3).

Table 32.3 shows the result of evaluation metrics (MAE) for recommendation system. It shows the improvement in the result of evaluation rates for proposal system comparing with previous system (k-means). At the Table 32.4, the proposal system is better than the previous system from level 2 to level 4 in the part of large purchase count. As a result of that, the performance of the proposal system is improved better than previous system from level 2 to level 4 although it is not good on level 1 and level 5 in the part of small purchase count. Figure 32.4 is shown in the site of recommendation of cosmetics on a smart phone. This system can be used immediately in e-Commerce under ubiquitous computing environment which is required by real time accessibility and agility because of finishing particular tasks such as clustering and calculating the probability of preference for pre-processing to reduce the processing time. in real-time environment.

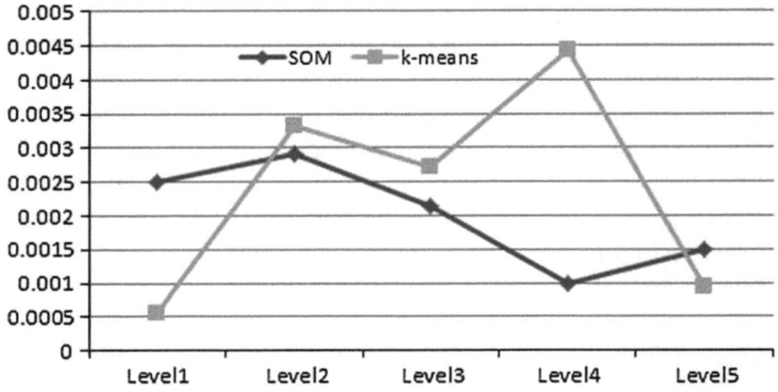

Fig. 32.3 The result for the graph of MAE by RFM score level

Table 32.3 The result of MAE by comparing proposal system with existing system

	P_count	Proposal (SOM)	Previous (KCA)	Existing
MAE	50	0.23	0.47	0.65
	100	0.13	0.23	0.32
	300	0.04	0.07	0.08
	500	0.03	0.05	0.06

Table 32.4 The result of MAE by RFM score level

RFM score level	SOM output count	k_means output count	SOM MAE	k-means MAE
1	52	18	0.0025	0.00056
2	24	12	0.00292	0.00333
3	158	77	0.00215	0.00273
4	524	280	0.00099	0.00443
5	74	54	0.00149	0.00093

Fig. 32.4 The site of recommendation of cosmetics

32.5 Conclusion

Recently u-commerce as an application field under ubiquitous computing environment required by real time accessibility and agility, is in the limelight [3]. We proposed an SOM clustering method using user's features to classify profitable customer for recommender service in e-Commerce in order to improve the accuracy of recommendation with an immediate effect We have described that the performance of the proposal system with SOM clustering method using user's features is improved better than previous system (k-means) from level 2 to level 4 in the part of large purchase count although it is not good on level 1 and level 5 in the part of small purchase count. It could make appropriate recommendation for each user's level possible based on neural network in real-time environment. We could simulate the SOM Clustering Method using user's features to classify profitable customer, generate recommending items to be possible to measure the purchasability for the future [4]. Thus, we could make the level of RFM score for the measurement of accuracy and efficiency, and validate the system by our results, then we can recommend items by each user's level according to the loyalty of RFM factors. To verify improved better performance of proposing system, we carried out the experiments in the same dataset collected in a cosmetic internet shopping mall. It is meaningful to present an SOM Clustering Method using user's features to classify profitable customer for recommender service in e-Commerce. The following research will be looking for ways of a personalized recommendation using fuzzy clustering method to increase the efficiency and scalability.

Acknowledgments This is work was supported by funding of Namseoul University.

References

1. Cho YS, Moon SC, Ryu KH (2012) Mining association rules using RFM scoring method for personalized u-commerce recommendation system in emerging data. In: International conferences, SecTech, CA. CES3 2012, Held in Conjunction with GST 2012. Communications in computer and information science vol 341, pp 190–198
2. Cho YS, Moon SC, Jeong SP, Oh IB, Ryu KH (2012) Clustering method using item preference based on RFM for recommendation system in u-commerce. Ubiquitous Inform Technol Appl LNEE 214:353–362
3. Cho YS, Moon SC, Noh SC (2013) Weighted mining association rules based quantity item with RFM score for personalized u-commerce recommendation system. In: The 8th international conference GCP2013, LNCS, vol 7861, pp 367–375
4. Cho YS, Moon SC, Ryu KH (2014) Clustering analysis by customer feature based on SOM for predicting purchase pattern recommendation system. J Korea Soc Comput Inf 19(2)
5. Deboeck G, Kohonen T (1998) Visual explorations in finance with self-organizing maps. London, Springer
6. Hastie T, Tibshirani R, Friedman J (2001) The elements of statistical learning—data mining, inference, and prediction. Springer, Berlin
7. Kohonen T (2000) Self-organizing maps. Springer, Berlin
8. Wei J-T, Lin S-Y, Wu H-H (2010) The review of the application of RFM model. Afr J Bus Manage 4(19):4199–4206

Author Index

J.J. (Jong Hyuk) Park et al. (eds.), *Future Information Technology - II*,
Lecture Notes in Electrical Engineering 329,
DOI 10.1007/978-94-017-9558-6

MIX
Papier aus verantwortungsvollen Quellen
Paper from responsible sources
FSC® C105338

If you have any concerns about our products,
you can contact us on
ProductSafety@springernature.com

In case Publisher is established outside the EU,
the EU authorized representative is:
Springer Nature Customer Service Center GmbH
Europaplatz 3, 69115 Heidelberg, Germany

Printed by Libri Plureos GmbH
in Hamburg, Germany